# Keeping Geese

BREEDS AND MANAGEMENT

# Keeping Geese

## BREEDS AND MANAGEMENT

Chris Ashton

THE CROWOOD PRESS

First published in 2012 by
The Crowood Press Ltd
Ramsbury, Marlborough
Wiltshire SN8 2HR

enquiries@crowood.com

**www.crowood.com**

This impression 2024



**British Library Cataloguing-in-Publication Data**
A catalogue record for this book is available from the British Library.

ISBN 978 1 84797 336 8

Typeset by Servis Filmsetting Ltd, Stockport, Cheshire

Printed and bound in India by Thomson Press India Ltd

# Contents

# Preface

We encountered goose keeping by accident; nothing was planned. The hobby just grew from a present of two small goslings for my daughter's birthday. Our geese were never considered for the table, yet in general they do have many values and uses. This is what the book is primarily about.

One reputation was quickly dispelled. They are not the farmyard tyrants of myth and memory. We found them to be intelligent, responsive, sociable and highly absorbing, especially if reared from goslings. They offer a real window into the 'wisdom of birds': close contact with geese really does open up a new perspective on avian intelligence. This book tries to put the record straight.

From a purely practical point of view, geese are the ideal food converters in a large garden, excellent lawnmowers and fascinating pets. They do well on a diet consisting largely of grass, and are cheaper to keep than other poultry which need a higher proportion of grain in their diet – an important consideration in an overcrowded world.

# Acknowledgements

Limited written information was available in the UK about keeping and rearing geese when we first acquired these birds. This book and *Domestic Geese* (1999) were the result of shared experiences with other goose keepers over the last thirty years. Keeping, breeding and exhibiting these birds has led to an accumulation of information which I hope other keepers will find useful.

Special thanks are due to John Hall for his help and advice on the breeds, and to Tom Bartlett for his expertise and sheer enthusiasm for waterfowl at Folly Farm in the Cotswolds in the late 1900s. When we started with geese, these experts led the way.

Other goose breeders in the UK and abroad have also contributed to the experiences, knowledge and photographs in this book: Peter van den Bunder, Bart Poulmans, Sonja Sauwens and Gerard Lambrighs of the Flemish Waterfowl Association; Andrea Heesters and Peter Jacobs, Hans Ringnalda and Sigrid van Dort, plus Kenneth Broekman of Holland; Michael Peel, Hamish Russell and Dr Harry Cooper (Australia's favourite vet); Bob Hawes from the USA for his continued support and detailed knowledge of the history and genetics of domestic geese; also Lori Waters, who has a real eye for photography, with her Runners and Sebastopols, and a passion for the birds.

Visiting shows in Germany and judging in France, Holland and Australia has also provided me with an insight into how the goose breeds have travelled the world, and how they continue to fascinate as a hobby bird and subject for photography. I am also grateful to Paul-Erwin Oswald, Mark Hoppe and Helene Towers for helping with translations and queries regarding the long history of geese in Germany.

Last, but most of all, I am grateful for the involvement with the geese over the years from both my daughter, Tig, and husband, Mike Ashton, who also looked over the manuscript. Without family involvement, keeping a multiplicity of birds is almost impossible. These birds have been part of the family for such a long time, and have brought us into contact with so many like-minded friends.

# Part I
# History and Origins

## Introduction

The goose was domesticated early and so has an old Indo-European name. Modern names derive from the Indo-European word '*ghans*': this led to Old English *gōs*, plural *gēs*, German *Gans* and Old Norse *gas* (as in Skånegås). In Greek the word became *khen*, and in Latin, *anser*.

Unlike poultry which derive from the jungle fowl, domesticated geese are recognized in only a few colours and shapes. This is surprising in view of their long history of domestication, spanning hundreds, if not thousands of years in widely separated areas of the globe. This wide distribution is reflected in their exotic names. There is the Sebastopol from Eastern Europe with its white, curled feathers, the proud 'African' from the Far East, and the Pilgrim, first standardized in the USA. Yet despite this wide geographic distribution, all breeds and varieties of domestic geese probably originate from just two wild species: the Greylag goose (*Anser anser*) and the Asiatic Swan goose (*Anser cygnoides*).

Human selection has meant that the birds have gradually changed in form and colour from their wild ancestors, and there are now many different standardized breeds and commercial strains around the world. These are now identified in the database pertaining to the United Nations Food and Agriculture Organisation Animal Genetic Resources (FAO ANGR, *see* Chapter 7).

# 1 The Origins of the Domestic Goose

## Early Domestication

Charles Darwin considered there was archaeological evidence for the domestication of geese in Egypt more than 4,000 years ago, since Egyptian frescoes depict cranes, ducks and geese, recording their fattening and slaughter. The fresco of the six geese at Meidum shows Greylags, White-fronted and Red-breasted geese: now at the Cairo Museum, the painted plaster is dated as the Fourth Dynasty, 2575–2355BC. The emphasis on geese is not surprising. The sacred animals of Amun, Lord of Creation, were the ram and the goose, kept at temples throughout Egypt, and although it cannot be certain that the Greylags were tame because paintings do depict the trapping of birds in the marshes, domestication would soon follow from the natural behaviour of the birds. They are thought to have been fully domesticated by the time of the Egyptian New Kingdom (1552–1151BC): for example, a flock of geese are counted and caught in *Nebamun's Geese* at Thebes, painted in about 1350BC (British Museum).

There are other species of wild geese which could have been domesticated: the Bar-headed goose (*Anser indicus*) from India and Central Asia, the Bean goose, White-fronted and Pink-footed goose from Europe, and the Canada (*Branta canadansis*) from North America. In some of these, smaller size was a limiting factor, and perhaps also behavioural characteristics; Egyptian geese (*Alopochen aegyptiacus*), for example, tend to be aggressive.

In China, the wild Swan goose was the foundation species for the domesticated Asiatic breeds, which include the 'Hong Kong goose' (African) and the Chinese. The breeding range of the wild bird was formerly much more extensive than it is today, stretching from Japan and Korea into the interior of China, Mongolia and Russia. The wide range of the bird means that it could have been domesticated almost anywhere in East Asia, but the origin of the Chinese and 'Africans' is generally taken to be China. Delacour states that the goose was raised for many centuries in China, beginning over 3,000 years ago. Pottery models of ducks and geese dating back to about 2500BC suggest early domestication (Clayton, 1984).

*Bean geese: it has been suggested that the domestication of the goose in China might have involved the Bean goose, in addition to the Greylag and Swan goose. Analysis of the karyotype of Chinese and African geese suggests otherwise (*see *Chapter 8**).*

*Pink-footed goose (*Anser brachyrhynchus*). This species breeds in Greenland, Iceland and Spitzbergen, and overwinters in Britain and mainland Europe. It is not as large as the Greylag, and is more wary in the wild.*

*The Swan goose, ancestor of the domestic Chinese goose and African goose.*

*The Eastern Greylag is a slightly larger and paler grey bird than the Western Greylag, with a slightly longer, pink bill – Sauwens-Lambrighs.*

## European Geese

Ancient domesticated breeds, such as the Roman, are acknowledged to have been selected and developed from the Greylag. The species was once common throughout Europe and Asia, overwintering in North Africa, Greece and Turkey, as well as India, Burma and China (Owen 1977). There are two 'races', which are slightly different: the Western race is more orange in the bill, whereas Eastern birds are bigger, have slightly paler plumage and an attractive pink bill. Todd (1979) attributes this colour difference to natural selection in two once-distinct populations that were divided by the Pleistocene ice sheet. But as the ice retreated, the populations merged again so that today there are gradations of feather and bill colour across Europe.

The Greylag has several points favouring its selection for domestication. First, it did not always migrate, but 'lagged' behind, often failing to follow the truly migratory Pink-footed and White-fronted geese north to their breeding grounds. Second, its range is extensive, which offers numerous geographical possibilities for its domestication. It also has a particularly amenable temperament: it is more tolerant of disturbance by man than many others, it is the most adaptable of European geese in food requirements, and will nest closer to man than other species. One population breeds in Iceland and overwinters in Britain, whilst another group breeds in Scandinavia and central Europe and overwinters in southern Europe and North Africa.

It also used to breed in the East Anglian Fens and the Netherlands before the destruction of the marshes. Weir (1902), for example, quotes Pennant from 1776:

> This species resides in the fens the whole of the year: breeds there, and hatches about eight or nine young, which are often taken, easily made tame, and esteemed most excellent meat, superior to the domestic goose.

As a gosling, the Greylag readily imprints on humans, and for this reason became the subject of research on goose behaviour by Konrad Lorenz. As a child he had been fascinated by geese and returned to them for material for his studies. Those who have kept and reared both wild and domestic geese will find it easy to understand why early subsistence people in Europe would have learned to domesticate these amenable birds and to develop their size on the farm.

More details of domestication emerge with written records, such as Homer's reference in the *Iliad* to Penelope's tame geese at Ithaca during the Trojan War (regarded as the twelfth century BC): 'I have twenty geese at home, that eat wheat out of water, and I am delighted to look at them.' At that time, according to Harrison Weir, Homer did not mention the hen.

By the fourth century BC the Romans kept a white goose developed from the Greylag. Lucretius wrote: 'The white goose, the preserver of the citadel of Romulus, perceives at a great distance the odour of the human race', and Virgil also ascribed the preservation of the Capitol to the 'silver goose': this is because in 390BC these sacred white geese saved the Capitol by their cackling warning about invaders:

> The Gauls . . . climbed to the summit in such silence that they not only escaped the notice of the guards, but did not even alarm the dogs,

> animals particularly watchful with regard to any noise at night. They were not unperceived however, by some geese, which being sacred to Juno, the people had spared, even in the present great scarcity of food; a circumstance to which they owed their preservation, for the cackling of these creatures, and the clapping of their wings, Marcus Manlius was roused from sleep . . . and snatching up his arms, and at the same time calling to the rest to do the same, he hastened to the spot, where, while some ran about in confusion, . . . he tumbled down a Gaul who had already got a footing on the summit.
>
> (Translation from Livy, AD59–17)

In *Roman Farm Management* by Marcus Porcius Cato (ed. F. H. Belvoir, 1918), writers from the second and third centuries BC gave accounts of goose keeping in Italy. The slave was recommended to select geese of good size and white plumage, because those with variegated plumage, which are called 'wild', were only domesticated with difficulty. Furthermore rearing goslings came with advice that would be equally appropriate to smallholders today:

> . . . it is not expedient to assign more that twenty goslings to each goose pen; nor on the other hand must they be shut up at all with such as are older than themselves, because the stronger kills the weaker. The cells wherein they lie must be exceedingly dry.

*Western Greylags at the Wildfowl and Wetlands Trust, Slimbridge, Gloucestershire.*

Columella, writing in the first century AD, also stressed that white birds were desirable. He described the fattening of the birds on crushed barley and wheat flour over a period of about two months, in much the same way as reported by Cato. Similarly Pliny Secundus (*c.* AD40, quoted in Harrison Weir) pointed out that, in Italy, white plume and down were preferred to the grey:

> A second commoditee that geese yield (especially those that be white) is their plume and down . . . the finest and the best is that which is brought out of Germanie. The geese there be all white and truly a pound of such feathers is 5 deniers [3 shillings and 1 penny] . . . many complaints are made of officers over companies of auxiliary soldiers . . . they license many times whole bands to straggle abroad, to hunt and chase geese for their feathers and down [the wild goose chase, Weir suggests].

Even in Pliny's time, the North European plain was producing the forerunners of the white Embden geese of today. Flocks of the birds travelled on foot from the north of France to Rome: 'Those which are tired are carried to the front, so that the rest push them on by natural crowding. . . . In some places they are plucked twice a year.'

Pliny also remarked that 'our people are aware of the goodness of their liver. In those that are crammed it increases to a great size . . .' thus indicating that the practice of producing *pâté de foie gras* is an ancient mistreatment.

Despite the relative wealth of evidence of the goose in Roman culture, Italy is not noted for goose production today – indeed nor may it have been then. The goose is a central and north European bird, and references to the birds from Germany and France indicate that they were probably more numerous there, and better developed from the wild Greylag stock that was captured from the wild breeding population. It is just that the early written records were – and still are – available from Rome.

## Goose Culture in Britain

It is quite likely that the goose was domesticated in Celtic cultures even earlier than in Rome. The bones of Greylag geese have been found in archaeological digs in the UK, but were they wild or semi-wild? Robinson (1924) records that when the Romans first invaded Britain in the first century BC they found tame geese, for Caesar reported that they were kept only for 'sport'. Greylags are thought to have been part of the stock at Iron Age dwelling sites. There is a suggestion that the natives appeared to treat geese as sacred, but this situation cannot have continued; with domestication, the farmyard goose became part of the rural economy.

Whatever the geese were kept for, there is a great gap in the records between Roman and medieval times, due to the lack of documentation. Robinson suggests that both the Anglo-Saxon invaders and the Normans could have introduced geese – or the custom of keeping geese – to Britain, since these invading groups came from goose-rearing lands.

Harrison Weir researched manuscripts from religious establishments and leases and discovered the ubiquitous goose in everyday life after the Norman Conquest. The goose was involved in paying for the tenure of land; for example, Weir cites a reference to William of Aylesbury, who held land under William the Conqueror: apart from straw for the bed, both eels in winter and green geese in summer had to be provided for a king's visit. Weir maintains that until the reign of Edward III that tenure was usually paid in kind, and in the list of requirements, Michaelmas goose was often specified.

In the section 'Fairs, Qualities, Usefulness and Value' Bishop Greathead's [Grosseteste] *Tratyse on Husbandry* from the thirteenth century records that: 'Ghees and hennes shall be at the deliverance of your baylyfe for lete, so ferme a goose for XII*d*. in a yere' and that 'evry goos shall answere you of VI ghoslynges'. In 1377 the market price of a goose in the city of London was 6d, whilst the wild mallard was 3d ('d' being old pence – 240 to the pound Sterling).

How good were geese to eat? There were varying reports: Weir quotes Andrew Borde (1542) as not praising the flesh of duck or goose 'except in a young grene goose', and also says 'the tame gese be hevy in fleinge [flying], gredi at their mete, and diligent to theyr rest – and they crye the houres of ye night – gose fleeshe is very grose of nature in degestion.'

Green geese were customarily young geese around ten weeks old, perhaps fattened for a short time on 'Skeg-Oats boyled' according to Gervase Markham (1613). Older geese of five to six months, which had gleaned the stubble and eaten the fallen grain, were later also fattened with oats, beans and barley.

Geese at that time did not generally seem to be very productive: the number of goslings they produced was limited – perhaps because the goose was allowed to sit, restricting her to one clutch a year – and geese were plucked for their down, which would also have limited their production. Perhaps their greatest value was in being a simple sideline to the farm: with a bit of care they would feed and look after themselves for most of the year, and more importantly, would also provide goose fat, feathers and down. In 1417 Edward V is said to have ordered that six wing feathers were plucked from every goose for arrows, and paid to the king. The quills were also used for writing. In both cases (arrow and pen) it was the outer flight feathers that were needed.

Did anyone pay much attention to the type and colour of their goose? Probably not: a goose is a goose, if it is to be eaten. Some of the geese may have been white, as mentioned by Markham in the 1600s, but many were grey:

> The large grey Goose that is bred in the Fen Countries is preferable to any other Kind, both for Flesh and Feather, and it grows to the biggest Size of any. We have besides this a smaller grey Goose, and a small dark colour'd [if which kind thre are some almost black]. Neither of these are nearly so advantageous as the grey; and among the large Kind that are so call'd, those are much better which are all of a Colour, than such as pyed or mottled.
>
> (Thomas Hale, 1756)

As well as on smallholdings and farms, geese were also produced by the cottager or commoner who had rights to graze on the moor or common – this was thought to be the origin of the name 'common goose'. It was better that the geese were not fed on greens and grain in a poultry yard because they would soon cost more to keep than they would fetch. So when the goslings were old enough, they were trained to go out in flocks with the goose and gander to forage over moor, meadow and common. Where many people kept geese and the grazing birds mingled during the day, various means were used to distinguish ownership, such as marks on the webs, coloured ribbon tied to the wings, or leg rings. But usually it was the tightly knit groups of the parent birds and their offspring which sorted out themselves in the evening. Where the birds were sufficiently well trained to move out to graze and return by themselves at the end of the day, it spared the expense of the 'gozzard' to drive and look after them.

As England's human population grew, the demand for food in town and city began to affect methods of food production and transport from the countryside. The development of the Aylesbury duck industry and the supply of ducklings to the London market are both well documented, but a goose-rearing system from the 1700s that is akin to the duck-rearing industry, is quoted from Pennant:

> A single person will keep a thousand old geese, each of which will rear seven . . . . During the

> breeding season these birds are lodged in the same houses as the inhabitants, even in the very bed chambers; in every apartment there are three rows of coarse wicker pens, placed one above the other; each bird has its separate lodge, divided from the others, which it keeps possession of during the time of sitting. A person called a gozzard i.e. a goose herd, attends the stock, and twice a day drives the whole to water; then brings them back to their habitations, helping those that live in the upper storeys to their nests without even misplacing a single bird. Vast numbers are driven annually to London.

Geese had long been caught and reared in the Fens, and by the late 1700s droves of domestic geese, often in flocks 1,000–2,000 strong, wended their way on foot to the capital. In Harrison Weir's account, these birds could cover more that eight miles (13km) a day, starting at three or four in the morning and walking until eight or nine at night. These large flocks were controlled by boys, each with a long hazel stick with a red rag at the end, and the gozzard who carried a crook to catch and select the geese when needed. Some birds might be sold on the way, others might need the 'hospital cart' when lame.

To protect their feet on these long marches they were driven through tar, which stuck to the feet, and then sand: this acted as a 'shoe' for the duration of the journey.

> Most of the flocks are the largest about harvest time when they travel from county to county in the south of England, being bought by the farmers to turn into the wheat, oat, rye or barley stubble after the corn is carried, when in a few weeks they are sufficiently fat for the poulter or higgler.

In addition to their meat, goose grease was good for chapped lips, and the internal fat made good pastry and was eaten on toast. The locks of guns were greased with it, as well as the wheelbox on the farm cart; also the scythe, and the steel knives and forks before being put away from daily use. This fat was considered of more value that hog's lard or any other farm product, grease or fat.

## The Twentieth Century and Today

During the twentieth century, agriculture changed immensely. Geese used to be found on every smallholding, but those smallholdings have now mostly gone. Few farms keep geese, and goose enterprises rearing a thousand birds or more are a rarity in Britain, although they do exist in Europe and China. Reports of Russian and Chinese goose production document traditional breeds and also commercial crosses for industrial scale production (*see* Chapter 7).

In Britain, goose numbers declined with urbanization, and the bird does not adapt easily to mass production. Yet it seems that over the last thirty-five years, the numbers of 'backyard' geese have increased, and interest in smallholdings, organic farming, food security and appropriate small animals for a mixed farm economy has grown. The goose fits the bill, mixing well with sheep and providing a good income at Christmas if marketing is well organized.

Yet for many, Christmas eating is not the main purpose of keeping geese. As with rare breeds of poultry, pigs and sheep, the birds are kept primarily out of interest. Despite Charles Darwin's comment that 'no one makes a pet of a goose', they *are* now pets, which mow the lawn, guard the house, go to the shows and amuse visitors. And for the pure breed enthusiast there is the pleasure of keeping perhaps a rare breed, and hatching and rearing each year's new set of goslings.

*A grey-and-white goose with a white gander exhibited in 1854. The early waterfowl exhibitions had a category for the 'common goose' which was, of course, the goose kept on common land. In parts of Europe, including Britain and France, a preference seems to have evolved for a gander that was invariably white '. . . even if the geese are grey'. It is not known when the habit started, but such selection resulted in auto-sexing breeds.*

# Part II
# Setting Up

## Introduction

Of all domesticated animals and birds, geese are probably the most diverse in the products and services they can provide. That is why they were on many subsistence smallholdings in the past: they are economical to feed, and are very hardy as long as they are protected from predators; they can help manage the land, provide useful products for the table, and are worth their weight in gold in sheer entertainment value.

If you want to keep geese, do find out about their intrinsic behaviour first, because their natural habits make them a very special bird. People keep them because they love them – and that applies to commercial keepers in Britain, too. Raised well, they are sociable companions, quite unlike the reported image of the Roman goose.

People are often quite unaware of the variety in size, shape, colour, temperament and habits of geese. So whilst running through the reasons for keeping these birds, it's also worth reading about the characteristics of the individual breeds before making a choice.

# 2 Why Keep Geese?

## Geese as Pets

Our own geese are kept for exhibition purposes and breeding, and are all regarded as pets. Well reared, geese stay tame for life; they are easier to handle than nervous birds, and are what people want to buy. Tame birds which come to you are simply easier to look after, and far more enjoyable to keep, than birds which run away. Geese are individuals, have more brains than chickens and ducks, and are great characters – and each bird does have its own personality.

Temperament to a certain extent goes with a breed. Africans are usually the calmest and tamest of all, but can be imposing because of their size and voice. Brecons should have a calm disposition too, but there are strains that disprove the rule. Whatever the generalization about the breed, it is the strain and the bird's upbringing which counts the most.

However, exhibition pure breeds and commercial geese are very different from each other. Commercial geese are usually white, and there are many varieties which have been developed for the table. They are often described as 'Embden', which they are not. The larger commercial strains are crossed with Embdens, but do not grow to the size and stature of the exhibition birds.

*Some breeds are difficult to get hold of, for example if they are rare, or only recently standardized – as these blue Franconian geese from Germany, belonging to Kenneth Broekman. Peter Jacobs*

White geese are generally preferred as table birds, and their eggs do hatch more easily than those of pure breeds such as the Brecon and Toulouse – but the advantages end there. In our experience, and with others we have talked to, commercial strains tend to be independent, self-sufficient birds. They have to be: they are reared in large flocks and have limited human contact. The ganders are often not suitable as pets, especially with children, since even hand-reared goslings change their characteristics quite markedly as they mature. Also, they are of no interest as a pure breed or show bird.

By contrast, pure breeds are reared in smaller numbers and the birds have generally had closer contact with humans, who probably, quite unwittingly, select the most amenable birds to keep. Breeding pure breeds rather than commercial geese may give lower production rates, but it gives added interest in many different ways. However, the most important factor regarding behaviour is the way the birds are brought up and handled. Sociable birds that you can enjoy come from a breeder who has spent time with them. Early familiarity with people determines their behaviour for life. If goslings are handled frequently, and are brought up by tame parents (goose or human), they will stay tame. By contrast, goslings with wary, aggressive parents, and badly handled by people, will also be aggressive, and it can be difficult to persuade them otherwise.

## The World's Best Lawnmowers

Geese are light-footed, useful grazing animals, and less likely to cause damage to land, fencing and greenhouses than goats and sheep which are so much heavier. They can live in the house and garden area, and are easily controlled with light fencing.

*These hand-reared Chinese goslings will happily approach people, and will stay tame for life.*

All geese graze copious amounts of grass, and if people realized just how much grass they ate, then the lawn-mower companies would have a huge drop in sales. Pilgrims are the best breed of all: they have not forgotten they are farmyard geese, whereas Toulouse geese still like their pellets. Goslings eat huge amounts of grass in order to meet their rapid growth rate, and will considerably reduce the need to mow, depending on the rate of stocking.

Geese will therefore save a lot of work on grass maintenance, especially in difficult corners and on steep slopes. They are excellent for grazing orchards, and in one particular case an owner chose them to maintain his Christmas tree plantation.

Grass grazed regularly by geese improves in quality – unlike the probing beaks of ducks which leave bare earth, thereby encouraging the growth of weeds. Geese will seek out all the dandelions, and a few young birds will even eat plantain. If there is a large infestation of these weeds, young goslings are best, and dense stocking of the birds for a short time is most effective. Chinese geese have slightly different habits from the Greylag-derived geese: they particularly like to root up creeping stems. Nettles and docks will remain and the birds will not touch thistles, but if the young plants are broken at the base, the geese will eat down into the root and kill them.

## Weeder Geese

Geese are selective as regards what plants they eat, preferring grasses to broad-leaved plants, and they can be used to weed grasses from a variety of crops. Goslings over six weeks of age are best, because of their voracious appetite. They will, of course, need protection from predators (such as dogs and foxes), to be watered well, and to be given supplementary food at the end of the day before being safely shut up overnight. Their great advantage is that they can be integrated into an organic system where there are no hazards from toxic substances – seed dressings, sprays and fertilizer granules. Young birds must be confined by light fencing to the area they are dealing with in order to supervise them more closely, but adult geese can be allowed to wander further away.

Chinese and 'Cotton Patch' breeds (*see* Chapter 11) were commonly employed as weeder geese in the USA: they are lightweight, active and relatively cheap to buy, and weed out perennial grasses. They have been used in vineyards, orchards and citrus groves, and also in crops such as coffee, cotton, mint, tobacco, hops and corn (after harvesting). Geese might be useful too, for weeding strawberries – they will, however, also eat the fruit, so this task would have to be early in the season and after the crop has been harvested. Always do a pilot study first to see how they behave before investing on a large scale.

In a garden, geese sometimes chew your best flowers, vegetables and saplings. Certain plants are toxic for them, and they can also be quite destructive – Brecons and Embdens in particular will chew twigs and ringbark saplings, so these need protecting with a tube of wire netting (plastic spiral may not be strong enough) wrapped around the trunk until the trees are mature. Greens such as lettuces, brassicas and even leeks will also be snapped up, so if the vegetable garden is important the weeding must be kept until after the crop is harvested.

If the main purpose for keeping geese is as pets to add interest to the garden and to weed the crazy paving and surrounding banks, lanes and yards, then Chinas are probably the best. They rarely chew bushes but may dig up patches on the lawn to dunk mud and roots into their bucket of water.

*Just two goose eggs fill this 30cm (12in) plate. The yolk is naturally orange from grass in the bird's diet, and the eggs are much healthier to eat than those from corn-fed birds. The albumen and yolk stand up well in fresh eggs. Note the small, pale germinal disc from which an embryo would grow in a fertile egg.*

## Guard Geese

Geese are watchful and inquisitive birds: they notice new people and unusual behaviour, and make a good deal of noise about it. Perhaps because of this reputation, and especially as a flock, they can deter people from entering premises – although this does not apply if the intruders have come to steal the geese! For example, white Chinese are employed as guard geese at a certain distillery warehouse in Scotland, where they patrol the fox-proof grounds at night. The idea is that the geese alert the watchman if there are any intruders – though ironically the geese were apparently once stolen. Known as the 'Scotch Watch', the flock now includes Romans, also famous in their own right in the history of the Capitol (*see* Chapter 1).

The noisiest birds are the best alarm, and this would be the Chinese and the African. But the most effective physical deterrent is perhaps a flock of white commercial geese.

## Geese for Goose Eggs

### Eggs for Eating

If large quantities of eggs are required, a utility strain of the Roman or Chinese breed is best, although the eggs of these small breeds are also smaller.

The yolk colour of the eggs from geese that graze is always a rich orange, and it is quite unnecessary to give a colour enhancer (*see* Chapter 16) in the diet. This is because the carotenoids lutein and zeaxanthin are obtained primarily from dark green, leafy vegetables, and these are responsible for making the yolk a deep orange colour. These substances are also thought to help prevent age-related vision loss. In addition, contrary to the fear that cholesterol in eggs was bad for human health, carotenoids also appear to inhibit the development of atherosclerosis by lowering the rate of arterial thickening. Furthermore the more natural the goose diet, the better the food quality of the egg. Thus poultry fed mostly grain produce egg yolks rich in

*Coffee cake made from a goose egg and wholemeal flour. (For recipes,* see *Useful Addresses.)*

saturated and mono-unsaturated fatty acids, whereas birds which feed on vegetation produce eggs with higher levels of polyunsaturated fatty acids.

Waterfowl eggs contain less water than chicken eggs and the protein content is higher, so when cracked on to a plate, the yolk sits up prominently and the albumen forms a neat disc instead of running away. It is this property which makes the eggs so successful for baking – better than chicken eggs; they also make the best custards and omelettes, though chicken eggs make better meringue.

### Eggs for Painting and Decorating

If egg size is important, then keep the larger goose breeds.

Eggs required for painting must first be blown – that is, emptied of their contents. This is done by piercing a larger hole in the rounded end and a smaller one at the pointed end; the contents are then broken up with a knitting needle, and blown out by air pressure. This can be done orally, or with a bicycle pump (using the attachment that blows up a football); pumps are also available for this specific purpose from egg decoration suppliers. However, the largest double-yolked eggs are not very good for 'blowing' as the shell is often thin and breaks into radial cracks when it is punctured

*Intricately carved and decorated egg, designed by Jackie Jarvis.*

*An egg which has simply been sawn in half and then decorated; design by Jackie Jarvis.*

After removing the contents, the eggs should be thoroughly washed out with bleach and then dried.

The advantage of the strong shell of the goose egg is that it can also be carved into intricate designs. Special cutting equipment is needed, and advice on equipment could be sought from the Egg Crafters Guild of Great Britain. This is definitely not a young child's activity.

To keep things simple, the eggs can be painted, or transfers (from egg decoration suppliers) used. Découpage – the art of sticking paper cut-out shapes to cover the egg shell – is also an effective way of decorating eggs, and the end result can be varnished to preserve the design.

If eggs are required for both hatching and painting, then 'incubator clears' can be used for painting: after five days in the incubator, and with the use of a candling lamp, the fertile eggs can be distinguished from the infertile ones. The latter will not be smelly, and will look little different in content from a non-incubated egg.

Eggs can be exhibited in the painted or decorated egg section at shows. Painted eggs simply have paint applied to the surface, whereas decorated eggs have various embellishments stuck to them as well. There are generally adult and children's sections.

Eggs are also exhibited fresh and judged for quality of both shell and content. The laying season for the goose is quite short as compared with the duck or hen, so it is often not possible to exhibit goose eggs. If the show does coincide with the laying season, waterfowl eggs invariably do well where the egg is judged for content because the yolk looks so rich and the albumen stands high.

## Geese for Exhibition

According to Ambrose (1981), the idea of exhibiting poultry first occurred in the early nineteenth century when village poultry shows became popular. However, as far as Edward Brown (1930) was able to trace, the first purely poultry show was held at the Zoological Gardens in London in 1845, with goose classes for 'Common Geese', 'Asiatic or Knobbed Geese' and 'Any Other Variety'.

*An amenable African gander at a show during a photography session.*

The first Birmingham Poultry Show was held in 1848, just a year before cock fighting was banned (1849). The inclusion of poultry in the Great Exhibition of 1851, and Queen Victoria's personal interest in exhibition poultry, also helped to increase its status, and further impetus was given to its popularity when the two most prominent agricultural societies in England – the Royal Agricultural Society and the Bath and West of England Society – each included a poultry section in their annual shows in 1853. This encouraged poultry shows to be held all over England, and led to their becoming a familiar part of agricultural life.

According to Harrison Weir, no geese were exhibited live for prizes before 1845. Ducks, chickens and geese were often shown dressed for the table, and even as late as 1954, a description of dressed poultry was included in the British Standards. However, the Victorian view of the exhibit had to change, since exotic poultry – imported at great expense and therefore valuable as breeding stock – were more interesting alive than dead. Historically, as long as contact between the different regions of the world had been limited by slow transport, exotic specimens had always been rare, usually arriving back in Britain skinned, stuffed or preserved. But the Victorian transport revolution had made it more likely that livestock would survive the long journey back from the Far East, and the expansion of the British Empire, together with faster transport, led to the export of British livestock to Australia and North America, and the import of plant and animal specimens to Kew Gardens and London Zoo.

As a result, where traders had been mostly limited to the English white Aylesbury duck and the common goose, they were now able to introduce the Toulouse, Sebastopol and Hong Kong goose, and this led to a tremendous increase in interest in these imported breeds. They were painted, exhibited and publicized in the press – but as they were increasingly judged alive, there was a need to establish a standard, describing the feathers, size and shape of each breed of bird.

The first *Standard of Excellence in Exhibition Poultry* was published in 1865 and included the Aylesbury, Rouen, Black East Indian and Decoy (Call) duck, together with Embden and Toulouse geese. Over the years, more breeds have been added as they have been imported or have become more numerous.

Until 1982, the indigenous geese of Britain (the 'common goose' of 1845) were neglected in favour of imported breeds – and even in 1982 it was the American standard for the Pilgrim which was used, since the Pilgrim was recognized as a breed in the USA whilst it was ignored in Europe. The *British*

*Waterfowl Standards* now gives recognition to our 'farmyard breeds', and is a very useful guide to the breeds; furthermore the 2008 edition has extensive colour photographs. Before buying or exhibiting any pure breed of goose, it is essential to be familiar with the standard.

Shows require a good deal of planning, so an entry form is provided for exhibitors to book and pay for the birds' pens in advance. The show provides a penning slip, which allocates pen numbers to exhibitors, either posted in advance, or given out on the day. The larger events produce a catalogue and have catering facilities and trade stands of interest to livestock owners. Some also have exhibits of other small animals, and all in all such shows make a very good family day out.

However, goose exhibits tend to be low in the UK: it is difficult to take many geese to a show in the usual family vehicle. Exhibitions on the continent are very much larger, and German shows have good line-ups of their speciality breeds such as the Embden, Steinbacher, Pomeranian and Elsässer (Alsace).

## Geese for the Table

Geese can, of course, be kept for their original utility purposes. On a well organized farm with appropriate facilities they are easy to look after because they are such an intelligent flock bird. Birds brought up together will stay as a flock. They are also creatures of habit and will follow a routine of dispersing to graze during the day, and returning home at night to protected areas where food is provided.

However, the production and marketing of table geese, and by-products such as feathers, down and goose fat, are both highly seasonal and specialized. It is essential to get good advice on all aspects of production and marketing before embarking on a commercial venture. Pure breeds are generally not suitable for commercial production: they do not reproduce in sufficiently large numbers, and advice will be needed on food conversion and the best strain. The British Goose Producers website gives information on commercial geese, and is an excellent organization to join for advice.

Goose meat, once thought to be rather fatty, does have a higher total fat content than chicken or turkey, but less than beef or lamb. There is also a relatively low proportion of saturated fat, and a higher proportion of mono-unsaturated and essential fatty acids. It is therefore healthy food.

The same cannot be said of *pâté de foie gras*, produced by the forced feeding of confined ducks and geese (known as 'gavage') so that the liver becomes enlarged by up to ten times its normal size – becoming, in fact, diseased. When extracts of such a liver have been fed experimentally to mice, they too have developed diseased organs (*see* Useful Contacts). *Foie gras* is not produced in the UK, and production is also outlawed in most of Europe – that forced feeding exists at all is a constant issue in animal welfare, and it is a disgrace that it still occurs in EU countries but against EU welfare regulations.

## Feathers and Down

Goose quills and down were an essential part of everyday life before the twentieth century. These natural products are still used today and are far more sophisticated than the synthetic materials now available: goose down as light as air, quills for calligraphy and feather vanes for fletching all out-perform their modern substitutes.

When the longbow was an essential part of warfare at the time of Crécy (1346) and Agincourt (1415), the goose played a crucial role in that archers would select goose flight feathers to 'fletch' their arrows. These flights could have come from the farmyard goose, though it is more likely they were taken from the Greylag. The vane of the goose flight feather is very tough and resilient, and even today, experienced archers find that goose feathers work far better than substitute plastic.

### Quill Pens

In the past, records were kept by inscribing on wax or soft clay tablets, and reed pens were used for writing on papyrus in ancient Egypt. However, some of the Dead Sea Scrolls are said to be written in quill pen, and the technique was introduced to Europe by around 700AD. By medieval times the goose quill was the preferred method for writing on more durable parchment. The best quality feathers were pulled from living birds,

*'A goose quill gentleman': this appellation was given to lawyers who often carried their pen stuck behind their ear when at their office. The saying 'A goose quill is more dangerous than a lion's claw' came about because of the legal connotations. The quill was used in 1787 to write and sign the Constitution of the United States of America, and other great documents such as Magna Carta.*

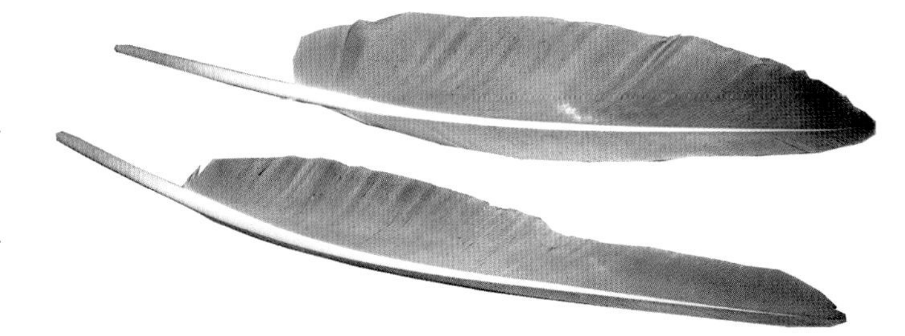

since the feathers and quills are not as strong once they are moulted, after a year's wear and tear.

The word 'pen' comes from the Latin word '*penna*', meaning feather. The favoured feathers for writing are the outer flights, which are more asymmetrical and also larger than the inner flights. For the best pens only the outer two or three feathers are used. The shaft of the feather acts as a reservoir for the ink. The quill was probably the instrument of choice for over a thousand years before it was supplanted by the metal dip pen, fountain pen and ball-point biro.

Goose quills can, of course, be collected each year from geese when they 'drop their flights' on moulting. Children love to use them in history projects at school; however, an adult should cut them into shape as the quills are amazingly tough.

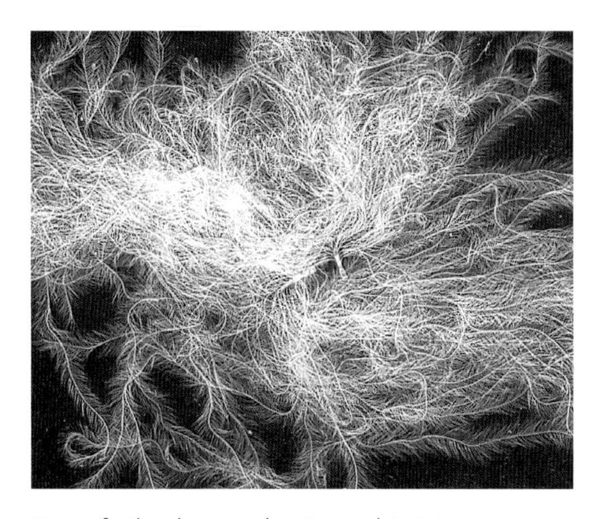

*Downy feathers have an almost non-existent stem. The long barbs have barbules, but they lack hooklets (barbicels) to lock the feather together. Down has the ability to expand from a compressed, stored state and to trap large amounts of insulating air.*

## Goose Down

The harvesting of feather and down was a necessity in the past. Geese were plucked of their breast down mainly in April when the feathers are supposed to be loose: this is when the broody goose plucks her own breast down for the nest, and even the gander's breast feathers start to fall. Harrison Weir (1902) quotes Pennant thus:

> It is for the sake of their quills and feathers that they are bred, being stripped while they are alive once a year for the quills and four or five times for their feathers. From this operation they do not in general suffer much, unless cold weather sets in, when great numbers perish in consequence.

The live plucking of geese (up to four times annually, for their down and feathers) is not permitted in the UK and most of Europe: it is a welfare issue because of the stress and physical damage it causes to the birds. An EU consultation document was produced on this issue in 2010 after the exposure of, and outrage at, such a practice. In agreement with European welfare regulations, the live plucking of geese should stop.

Goose feathers and down can be ethically produced from the dry plucking of dead table geese. This down is very clean, and just smells a bit 'goosey', like fresh grass. However, commercial goose feather and down products are advertised as 'steam cleaned'.

Unwashed down is very resistant to moulds and bacteria, and does not seem to be attacked by insects such as moths and beetles. Down in old duvets behaves as well as ever after even thirty years of use.

For the home preserving of down, gentle baking in the oven has been recommended, perhaps similar to the method described by Weir:

> The plan most generally pursued by farmers' wives . . . is to lay the feathers on paper on the floor of a spare room to dry, and then beat them lightly to get out the dust or dirt, after which the quill parts are carefully cut off, and then they are put into bags and placed in the oven after the baking, while it is yet slightly warm, and left there, when, after a few weeks, the bags are hung in rows on beams in an airy room so as to get thoroughly free from all moisture, and when that is so, are put into bed, pillow or cushion 'ticks' for use.
>
> (Weir, 1902)

Where sufficient quantity of feather and down is produced in the UK, goose producers market their own ethically produced feather and down products; alternatively the product can be processed by specialists located in East Anglia (*see* Useful Addresses).

# 3 How to Start

Welfare of both farm animals and pets comes first, so think twice about geese if space is limited, if neighbours are likely to complain about noise, or if your children are too young to cope with them. Geese are also a tie: they cannot just be left when you go on holiday - someone must look after them when you are away, and fox-proof fencing should be in place to keep them safe. Before making a decision on how many birds to keep, refer to the sections on understanding geese (*see* Chapters 4-6) and their requirements (Chapter 16). Also, it might be advisable to start off in a small way to see how you get on with them, and how the land responds.

Geese are usually sold in pairs, and are best kept in pairs unless high production is wanted. A gander often favours just one goose; females can pick on one another, and for peace and harmony, especially if the goose is sitting, a pair is easier to manage. Never shut up together birds that do not get on, because the stronger one can inflict serious damage on the other.

***Breeds of Geese in the UK Standards***

A breed is a group that has been selected by humans to possess a set of inherited characteristics that distinguishes it from other geese. A 'breed' must be recognized by a governing body. Breeders select individual birds from within the same gene pool to maintain the characteristics of the breed: that is, they choose individuals that will pass on desirable characteristics to their progeny. Breeders are generally familiar with standards and waterfowl exhibitions, and keep stock records.

| ***Heavy geese*** | *Weight in lb (kg)* | *Estimate of eggs per annum ** |
|---|---|---|
| African | 18-28 (8-13) | Up to 30 |
| American Buff | 20-28 (9-13) | 25-30 |
| Embden | 24-34 (11-15) | Up to 30 |
| Toulouse | 20-30 (9-14) | 30-40 |
| ***Medium geese*** | | |
| Brecon Buff | 14-20 (6-9) | Up to 30 |
| Buff Back and Grey Back | 16-22 (7-10) | 30 |
| Pomeranian | 16-24 (7-11) | 30-40 |
| West of England | 14-20 (6-9) | 20-30 |
| ***Light geese*** | | |
| Chinese | 8-12 (3-5) | Up to 40, or more ** |
| Bohemian [Czech] | 9-12 (4-5) | 40-60** |
| Pilgrim | 12-18 (5-8) | 25-35 |
| Roman | 10-14 (4-6) | 40-65** |
| Sebastopol | 10-16 (4-7) | 25-40 |
| Steinbacher | 11-15 (5-7) | 5-40 |

These breeds make good sitters - they are likely to go broody, and are not too heavy

** Estimate only. The number laid depends on the strain, the age of the bird, and how they are fed. Numbers quoted are for young birds, without artificial lighting*

*** The number very much depends on the strain. Geese mostly lay in spring only. A laying strain of commercial Chinese will lay up to eighty eggs, some of those in autumn*

**Heavy geese**

a brown African, male
b white African, male
c Toulouse, Male
d Embden, male
e American Buff, male

**Medium geese**

f Pomeranian grey back, female
g Brecon Buff, female

**Light geese**

h Pilgrim, female
i Steinbacher, male
j Roman, female
k Sebastopol, smooth-breasted male
l Sebastopol, curled-feather female
m white Chinese, male
n brown Chinese, male

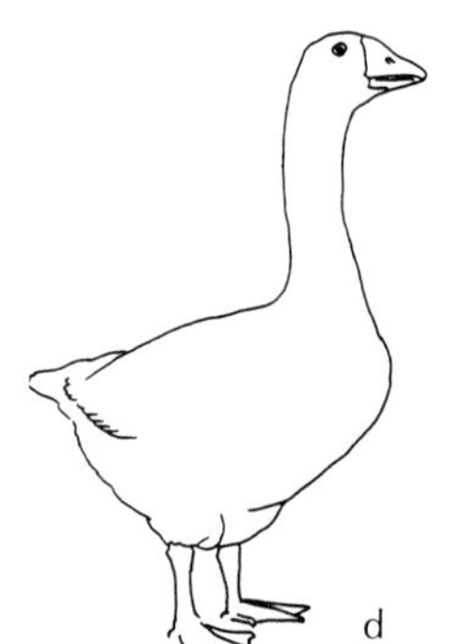

*Popular breeds of geese standardized in Europe.*

For that reason it is not advisable to introduce a new female to an established pair, because she will be disliked by the resident female. Geese do pair in the wild, and domestic geese tend to keep their natural habits.

It is important that the breed is fit for purpose. Pure breeds are no use for commercial table production because they don't hatch out in large enough numbers. Commercial producers have their own strains of white geese which have been selected to lay large numbers of easy-to-hatch eggs.

Check breed characteristics for size and behaviour. The heavier breeds require more space, larger housing and more washing water; they will also lay fewer eggs. The egg-laying strains are generally lighter in weight but can be noisy. They are, however, often more widely available because they are easier to produce.

Although the breeds do behave somewhat differently from each other (especially the African and the Chinese), strain and upbringing are the most important factors in goose behaviour – meet geese with their owners before you decide to buy. On the whole, well managed pure breeds make better pets, and are more interesting birds, than white commercial geese (*see* Chapter 21). Also, quite often an amenable white 'table' gosling can turn into the archetypal farmyard tyrant in the breeding season, and this situation is best avoided by good management and a sensible choice of bird.

*Waterfowl show: Normandy geese at Tours, in France, in November 2010. Colin Brierley*

## Getting to Know the Breeds

If the main aim is to produce a pure breed for exhibition, or for sale as a pure breed, the parent stock should conform to the published standard. It is advisable to check birds against the standards, and birds in breed classes at the larger shows.

The main waterfowl shows are in winter when the birds are in full feather, and when they travel better in the lower temperatures. To look at a specific breed, visit one of these where there is a class for every breed, and there may be several examples to see at first hand. Although the most accessible events are the summer County shows, these are frequently not well attended. Summer is the wrong time: it is often too hot for stock to travel, the birds are in moult, and breeders are too busy rearing youngstock.

It is also useful to visit an established breeder. Breeders are generally not open to the public because their geese are usually hobby birds. However, many of those who advertise are willing to show their birds to anyone with a genuine interest – though it is advisable to make an appointment first.

Birds unfortunately do not have pedigrees. In the absence of written records, the best insurance for a 'pedigree' bird is the closed breeder ring issued by each country in Europe (*see* Chapter 20). Specialist breeders of long standing will keep their own stock records and find closed ringing useful for this purpose. Such breeders are also likely to find new stock from abroad. This is important because closely related birds ultimately fail to breed. Geese suffer from a more limited gene pool than ducks because fewer birds are kept and it is much more difficult to keep the pure breeds going.

## Obtaining Stock

### Eggs and Goslings

Eggs of 'pure breed' geese are now commonly offered for sale. However, breeders with quality stock rarely sell eggs. Geese lay few eggs compared with ducks and poultry, often only between fifteen and forty. Fertility in goose eggs is also more random than in ducks, so these two factors mean that a hatch of eight goslings from one female is good for one season. Breeders who value their stock therefore need the eggs themselves, and where fertility is erratic, it is also difficult to arrive at a fair price.

Hatching eggs advertised in the paper, on sale at markets or on the internet, are unlikely to produce good examples of pure breeds. Vendors may or may not know the breeds, but you will not know how old the eggs are, how they have been stored, or even if they have already been incubated and are 'incubator clears' (*see* Chapter 19). Goose eggs also get very badly shaken up in travelling through the post because of their size, so fertility and hatchability tend to be poor.

Pure breed goslings are rarely sold by breeders unless the quality can be assessed early – as in the colour of Buff Backs and Pilgrims. Breeders aim to grow birds to see which are best, mainly because it is often impossible to tell at first glance which of the goslings will make the best birds: out of the same hatch there can be exhibition birds, some good quality breeders, and birds that are only pet quality. As a result they will vary in price. Breeders generally sell only well grown birds from August onwards. By contrast, commercial birds are hatched in higher numbers and should be available as goslings in the spring (*see* Chapter 21).

## Adult Birds

Adult birds are those which are in their second feathers and are at least sixteen weeks old. However, they are not sexually mature and they will continue to grow, especially the larger breeds.

Older birds, past their prime, are sometimes available as pet and garden stock, but guard against buying a female with a broken down, distended abdomen. Such birds often lay double-yolked eggs and are also prone to infertility.

For breeding, two- or three-year-old females are the best. Wild geese do not lay and breed until at least two or three years old. Domestic geese lay in their first year but, depending on the breed and strain, eggs from a yearling goose are not generally successful. She needs another year to mature before she will lay eggs of full size. Small eggs from a yearling goose may well hatch, and they can even hatch more easily than larger eggs from the mature goose, but the goslings are usually smaller and weaker and they will not produce good, breeder-quality birds.

Buying a two- or three-year old gander can also be advantageous. Sometimes a late-hatched male can fail to breed in the following spring. However, early hatched yearling ganders are preferred in the heavier breeds; they have not reached their full size and are more agile for breeding. Toulouse should be fit, not fat. Heavy breed males often have to be pensioned off by the time they are nine years old (or younger), whereas lighter weight breed ganders can still breed at twenty.

Ducks and drakes can be sexed easily by their plumage colour and voice, but geese are more difficult. Experienced breeders know the sex of the birds by their shape, voice and behaviour, but it is not a foolproof method. This is particularly so with the Toulouse, especially if hatches from different parents are run together in a flock and there turn out to be birds of different sizes for the same sex. Vent-sexing is the only certain method. Vent-sexing is much easier in goslings between three to four weeks of age when the birds are marked (*see* Chapter 20). Errors are most likely to be made with juvenile birds at eight to sixteen weeks, particularly with Toulouse which are a difficult shape to handle.

Some larger breeders have a price list, but it is very much easier to list a price for most poultry and ducks than geese. Geese do not lay lots of eggs, are not nearly as numerous as ducks, and their breeding is much more erratic even in the hands of the experienced. Geese are, however, long lived.

Neither goslings nor geese are normally 'shipped' in the UK. Professional carriers with a licence to carry livestock are hard to find, and stock is usually collected by buyers or delivered by agreement.

The larger autumn auctions sometimes offer pure breeds of geese, but problems can arise because the birds are bought 'as seen'. There can be good examples on sale, but quite often a breed listed under a pure breed name is a poor example, or not even the breed stated. In general, breeds of geese are not widely known, so it is not advisable to rely on other people's word or descriptions at auctions where there is also no guarantee of the age of the birds, or even of their sex. It has been known for a 'trio' to turn out to be three males. Only buy at an auction if there is reliable, independent advice. Most pure breeds are sold privately, as indeed are pure breeds of many other types of animal.

Note that pure breeds of geese have had time and care lavished on them by committed breeders for generations. Also, some breeds have only recently been imported. With small supply and relatively high demand, the price for some pure breeds will be relatively high, and this has always been the case with quality geese. In addition, breeders limit the number they produce to what they can reasonably rear in good conditions. Pure breeds are not mass produced.

## Choosing Quality Birds

If the main aim is to produce pure breeds for exhibition, or for sale as a pure breed, the parent stock should be good examples and conform to standard.

### Shape, Size and Colour

Birds are in their adult feathers by sixteen to eighteen weeks. By then the shape of a bird is well developed, though in the heavier breeds it is still not possible to determine how well they will eventually turn out. The keel of the Toulouse and the dual-lobed undercarriage and full weight of the Embden do not fully develop until the birds are over two years old, so young birds will have to be assessed for their potential.

Looking at parent birds will give a good idea of their offspring's potential. In some breeds, such as the Toulouse and Embden, size is very important and it is as well to choose big birds from the outset. In others, such as the Buff Back and the Brecon, colour and markings are of more importance, whilst in the Chinese it is their carriage and shape. Read the standard descriptions of the birds carefully and compare them with the individuals you intend to buy.

### General Faults

Apart from specific breed faults, there are general faults to look for and avoid when choosing stock. Never accept a bird already parcelled up in a box: it should always be picked up and its weight checked. If it is too light this could be caused by underfeeding or by worms, but it could also be ill.

Whilst holding the bird, also check its eyes. Are they the right colour for the breed, and are they healthy? Birds can have poor eyesight from an opaque growth over the pupil, and this will also affect their behaviour. Also check the bird's feet and legs. The hocks and ankles should not be hot or swollen. The toes should be straight, and in a young bird the undersides of the webs should not be thickened or callused.

Defects may not show up in young goslings, some of which may fail to grow to their full potential. Birds have to be at least sixteen weeks old before they can be assessed with certainty, and that is why goose breeders themselves buy well grown birds.

Birds should also be seen moving around. A bird's behaviour should be watched for any irregularities which may suggest poor eyesight. Structural deformity of the spine (curvature of the back or wry tail) may not be evident until a bird is half grown; deformity of the bill (undershot mandible) will probably show up earlier.

The wings need to be correctly set. Any gander with a slipped wing may not breed because he will find it difficult to balance in mating. For that reason, males should not have their outer flight feathers clipped. The majority of domesticated geese do not need wing clipping anyway.

Finally, the behaviour of the bird is important. Do the birds come to their owner, or run away? The behaviour of the birds is determined by both nature and nurture. Tame birds will soon adapt to a new home; nervous ones will need more time and care.

TIPS

* Plan your fencing, housing and grass before buying geese. Geese will end up depositing parcels of sloppy grass droppings by your back door if you let them.
* Geese and young children do not get on. Children must be old enough to understand that their behaviour will affect how the birds behave.
* Make sure any neighbours will not object to noise. Chinese can be noisy, and Africans are loud when they speak.
* Start off in a small way.
* Get to know the breeds and their behaviour.
* Make sure the geese conform to standard and are healthy.
* Ensure your birds are fit for purpose.
* Geese are a commitment. They are big birds and have a long life span (twenty years or more). Impulse buying is not recommended.
* Meet some geese and their owners before you buy. Geese are sensitive to body language. Is yours right for them?
* Only buy at auctions if you have the advice of an experienced person who can tell males from females, and can also assess the age, health and breed of the birds.
* Goslings should be purchased direct from the point of supply, not at auctions and sales, because of their welfare. Specialist facilities, food and protection will also be required for rearing them.
* Be aware that goslings may have acquired gizzard worm (if they have been out on grass) before purchase.

## Breeding Stock: Considerations

A 'breed' is a group of birds where the individuals are similar to each other in outward appearance. When bred together, the offspring should also be similar to each other – that is, the parents should 'breed true'. To achieve uniformity, however, a certain amount of inbreeding has to take place, and the following points should be borne in mind for quality stock birds:

**Inbreeding means that the gene pool of a group is restricted, therefore relatives which exhibit desirable traits are used in a breeding programme**. Pure breeds are thus produced by human selection of the offspring in this breeding process. This has resulted in specific breeds of cattle, pigs and sheep suited to differing physical and economic circumstances. The same process has also taken place with geese over hundreds of years. Inbreeding has both advantages and disadvantages.

**The purpose of inbreeding is to improve stock**, be it for aesthetic or practical purposes, for example to enhance characteristics such as size and colour. Thus there is no point in using inferior stock: the foundation stock must exhibit some particular characteristic which the breeder wishes to enhance. If, in a breeding pair, the characteristics of the gander are particularly desired, then he would be mated back to his best daughter showing these characteristics. It would be similarly so with a mother and son.

**This kind of mating can be used where the foundation stock was not closely related**: siblings and cousins can also be used instead of father/daughter, for example. Another strategy is to use grandparent and grandson/grand-daughter matings. Whatever the relationship used, records must be kept to monitor successes and problems.

**Inbreeding to maintain the phenotype inevitably reduces viability by restricting the gene pool**: this eventually results in any deleterious recessive genes ultimately being expressed in their homozygous form and showing up as deformities. Inbreeding therefore ultimately results in reduced fertility, leg weakness, blindness, spinal deformities and hereditary diseases. Clearly inbreeding which results in poor, weak stock should not be pursued.

**Birds with genetic defects must be avoided** so 'new blood' is added from time to time by selective out-crossing, or by the use of similar birds from a different gene pool in the same 'landrace'. In the case of popular breeds worldwide, new genetic material may be from imported stock.

**If the birds' breeding background is unknown, try to acquire those which are thought to be 'unrelated'.** It is desirable not to use siblings or a mother/son or father/daughter relationship unless the breeding record is known. Breeders can only carry out selective inbreeding if they know the stock's breeding record in detail, and how the geese are performing. Poultry breeders of years of experience consider that no more than four generations of inbreeding should be used before reverting to out-crossing to another individual of the same breed, but of a different strain. Note that some specialist breeders will keep more than one set of breeding geese, and can sell pairs which are not too closely related.

**Some breeds are in short supply because they come from a very few individuals.** There have been very few imports of Africans, for example. It is unrealistic to think that completely unrelated UK birds are available in show birds and rare breeds where a certain amount of inbreeding has to take place to retain the breed's characteristics. This is, after all, how pure breeds are made. In contrast, commercial geese are developed from a wide gene pool to ensure hatchability and vigour.

### BUYING PURE BREED STOCK

The advantages of buying pure breed adult stock from a long-established breeder are as follows:

* The breed should conform to the standard; it is not cross-bred.
* The birds are closed rung with their individual identity number.
* Advice about the breed's characteristics can be given.
* Other examples of the breed can be seen for comparison.
* The breeding background of the birds is known – that is, if are they siblings or unrelated.
* The age and sex of the birds are known.
* If there were a mistake, the bird/s can be returned.
* Advice about management can be sought.
* The birds can be handled and checked for health.
* Advice on health and worming can be sought.
* Examples of housing, feeding, bedding and fencing can be seen.
* The behaviour of the birds can be assessed.

# Part III
# Understanding Geese

## Introduction

Ducks, geese and swans belong to the family Anatidae. These waterfowl all have webbed feet, two layers of feathers, and an extra layer of fat under the skin to keep them warm. These are adaptations to their life on water, but the wild birds are also strong fliers. Most wild geese migrate according to the season to seek out safe places for nesting, and a good food supply. Due to their feeding habits they spend more time on land than ducks and are semi-aquatic.

Geese are flock birds and often fly in a skein in its characteristic V formation; they also 'roost' together (on water) at night. There is safety in numbers. However, within the flock there are strong and often lasting pair bonds. Family ties are very strong when goslings are raised, too, and the larger the family, the greater its territorial authority.

Feather maintenance and feather replacement (moulting) are timed to fit in with migration and breeding. The condition of their feathers and the stage of feather development both affect goose behaviour, too. Feather type and condition are crucial to their survival for warmth, flight, migration and nesting habits.

Geese are almost exclusively herbivorous, eating a wide variety of grasses and sedges, including seeds and roots. Their taste extends to cultivated crops where they can become a 'pest species' in large numbers. Occasionally geese will sample insects, but they are not regarded as omnivores. The bill, with its strong serrations (lamellae), is adapted to cutting vegetation, and the gizzard, with its addition of grit and sand, to converting cell sap into food supply.

Domesticated geese are very like their wild cousins. They have the same physiology and food requirements, and follow the same annual breeding and moulting patterns. Their social behaviour is also similar: the main difference is that they are much larger, less able to fly, and of course they no longer migrate.

# 4 Goose Anatomy

## The Head and Neck

The head of the goose combines many functions of the human arm because it acts both as a tactile organ and as a hand. Birds depend upon their bill to collect their food, build their nest, preen their plumage and defend themselves. Demonstrations with the head and neck are also used for communication in courtship, recognition and aggression (*see* Chapter 17).

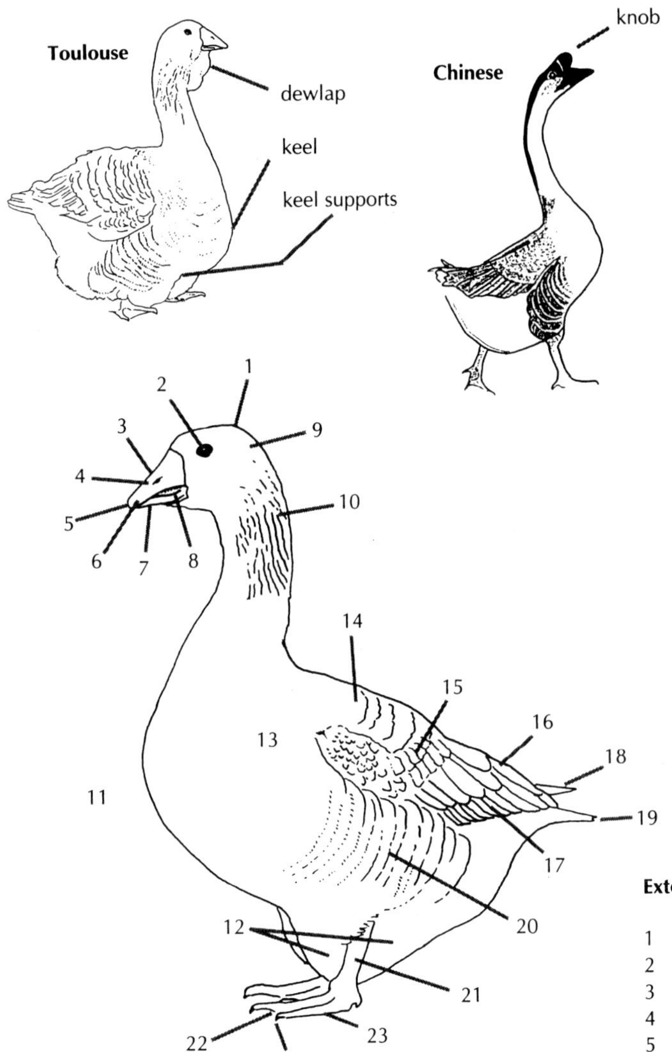


*Goose morphology.*

### The Bill and Mouth

The bill is the most obvious feature of the head, and has evolved as a light, strong covering of the jaws to replace the reptile-like heavier teeth seen in bird ancestors. Food processing is in two main steps: the bill tears the food, and the 'chewing' is now done by the internal gizzard. Tearing is therefore a rapid process, and taste is thought to be relatively unimportant, though geese do distinguish between nutritious, palatable foods and those of lower value. That distinction is tactile as well as visual.

The bill consists of an upper and a lower mandible. The lightweight bones of the bill are covered by a layer called the *mandibular rhamphotheca*, which resembles skin. However, this is thicker and harder because it contains calcium and keratin. The end of the bill (the *bean*) is hard, probably for protection whilst feeding. The rhamphotheca is leathery in waterfowl, but is thickened and modified at the edges, where it forms hard serrations for gripping and snipping grass.

The lower mandible is not solid: it has a fleshy central area in which the tongue sits. The upper bill, which is continuous, carries the nostrils (*nares*) which connect with the airways.

Inside the mouth are the tongue (which contains bone), the salivary glands, the opening to the oesophagus, and the opening to the windpipe (the *trachea*). On the upper palate is a slit (the *choana*) which forms a straight, closed connection from the nostrils to the windpipe when the bird's mouth is closed. The arrangement is different in mammals where air enters the back

**External goose morphology**

| | | | | | |
|---|---|---|---|---|---|
| 1 | crown | 9 | ear | 17 | secondari |
| 2 | eye | 10 | greylag neck-feather partings | 18 | primaries |
| 3 | culmen | 11 | breast | 19 | tail feathe |
| 4 | nostril | 12 | paunch or abdomen | 20 | thigh cove |
| 5 | bean (or nail) | 13 | wing butts or fronts | 21 | shanks |
| 6 | upper mandible | 14 | scapular feathers | 22 | webs |
| 7 | lower mandible | 15 | coverts | 23 | toes |
| 8 | serrations | 16 | tertiaries | 24 | claw |

*All the geese in this Sebastopol flock see a high-level aerial threat – except the baby, who is blissfully unaware of it. Lori Waters*

of the throat first, or enters directly through the mouth. At the base of the tongue is the *laryngeal mound* and the opening to the windpipe (the *glottis*). Unlike mammals, which use the larynx for vocalization, birds use the *syrinx*, located much further down the trachea.

### The Eyes

Avian eyes dominate the head. Freethy (1982) notes that 'A bird could almost be described as a pair of flying eyes' because birds have developed the largest eyes of any vertebrate group. Avian eyes are very perceptive: a large image is projected on to the retina, allowing a bird to see far more detail over a larger area than a human. When a bird scans the horizon, it perceives far more than you or I. Watch your geese: they react to specks in the sky that we barely perceive. They constantly scan the horizon, and assess detail and movement much faster than a mammal. Geese are, after all, a prey species, and need to be aware of the threat of predation, especially when rearing goslings.

Another difference between our eye and the bird's is that our lens filters out wavelengths of light below 400nm, making detection of ultraviolet radiation invisible. The avian lens appears to admit wavelengths of light down to about 350nm, which makes near ultraviolet radiation visible to the bird. The significance of this is that birds may secrete a substance from the preen gland that is spread on the feathers and is visible in the ultraviolet range. It is suspected that this may be one way that some birds visually discern the sex of other birds.

Bird eyelids are different, too. When they do close their eyes to sleep, their lower lid closes upwards and the paler surface of the lower lid seems to say that

*A three-week-old gosling, showing the beak serrations for snipping, and the nictitating membrane of the eye.*

they are still alert, not napping. They also retain a third eyelid, known as the *nictitating* membrane: this lies beneath the two main eyelids and is used for blinking and protecting the eye when in flight. The membrane is transparent so vision is not impaired.

Bird's eyes are not readily manipulated in the skull because they are quite large. This problem is compensated for, certainly in the goose, by the length of the neck and the manoeuvrability of the head. Geese will readily take an all-round view when assessing aerial threats. The neck also needs to be long enough to reach the preen gland to transfer the uropygeal gland oil over all the feathers. The vertebrae in the neck (generally more than twenty in long-necked birds) move freely, unlike those of the spinal column where some of the vertebrae are fused.

## The Body and Bone Structure

Birds still follow the general vertebrate plan, but they have important adaptations which have allowed them to fly. Not only have they reduced bone mass to make them much lighter, but they have also strengthened their framework by fusing bones for rigidity. As a consequence of bi-pedal walking and of flight, the centre of gravity of the bird has shifted, and the body is shorter than in four-legged animals.

## The Spine, Pelvis and Thorax

Widespread fusion of bones is seen along the avian spine. For example, three lumbar (abdominal) vertebrae are fused with the pelvic area and six of the tail vertebrae to form the *synsacrum.* This fused structure is a complete unit of bone, and supports the weight of the bird when it is on the ground. It gives a platform for the attachment of the leg and tail muscles.

In addition, the ribs are reinforced by *uncinate processes* – lateral braces attached to the rib. This strong network of bone and connecting ligaments resists the force of the breast muscles in flight. The breast muscles themselves are supported by a deep *sternum*, unlike the earlier birds which did not truly fly. The development of the sternum allowed the attachment of muscles to develop true flapping flight. Birds have come a long way from the simple gliding potential of their ancestral dinosaurs.

The lightweight bill (replacing heavy teeth), the reduction in the weight of bone, and the fusion of parts of the skeleton for strength, have all contributed to efficient flight. Many of the bones of the avian skeleton have air-filled cavities instead of bone marrow, these pneumatized bones connecting with the air sacs of the respiratory system.

## The Wings

The pectoral girdle anchors the muscles of the wings to the body. It is a 'free-floating' structure held in muscular supports. The *coracoid bone* connects the sternum to the shoulder joint where the *scapula* (like our shoulder blade) is also attached. The *clavicle* (equivalent to the collar bone) is the wishbone; it is an attachment site for the breast muscles. Where these three bones meet is the very important *trioseal canal* where a tendon, essential in raising the wings in flapping flight, sits in place.

Wings themselves can be more easily understood if they are compared to the human arm. The upper arm (*humerus*) is short and stout because flight muscles are attached, and it has to bear the stress of upstroke and downstroke. The *radius* and *ulna* are similar to our forearm, although after that, many of the 'hand' and 'finger' bones are fused.

The wing butt of the goose is equivalent to our wrist. Here, two *carpal bones* are thought to restrict the movement of the 'hand', and therefore keep the primary feathers in alignment in flight. The bones of the 'hand' are so extensively fused into the *carpo-metacarpus* that distinction of the fourth and fifth digits

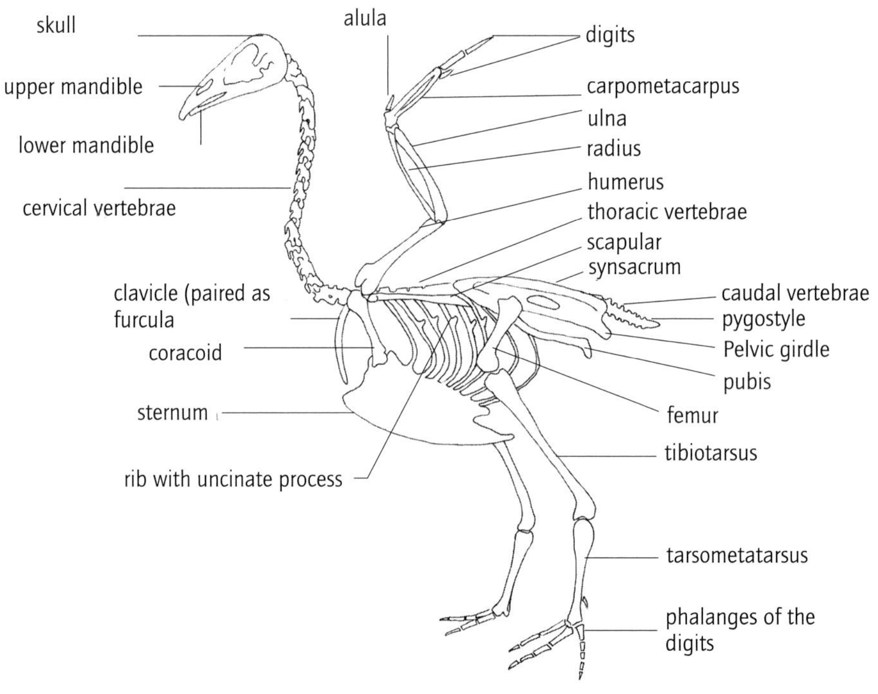


*Generalized goose skeleton based on the Greylag.*

has been lost. However, the first digit can be seen near the wrist joint and is called the thumb. It supports the *alula* feathers.

### The Legs and Toes

The basic structure of the legs is also similar to humans, but much of the leg is not seen. The short *femur* (thigh bone), knee joint and *tibiotarsus* (shin) are tucked into the feathers. The hock and shank of the bird is visible, but these are equivalent to the bones of our heel and foot, not the human shin bone. The shank – *tarsometatarsus* – is made from fused bones of the foot, and this design gives extra strength in pre-flight take-off and landing. Birds therefore walk on modified toes, and the digits are generally reduced to four. Unlike the fused 'fingers' in the wing, where the *phalanges* (finger bones) are no longer distinguishable, the phalanges in the toes do articulate.

The first digit, the *hallux*, is equivalent to our 'big toe', and points backwards. It is important in perching birds, where it becomes fixed in position for secure perching. Of course this faculty is not important in geese, yet this small toe still becomes fixed in position after five weeks of age.

A tough plating of scales strengthens the foot and protects it. Periodically a layer of dead scales falls off and reveals a new layer underneath. Geese and ducks have a webbed foot, referred to as palmate, where the anterior toes are connected by this scaly skin.

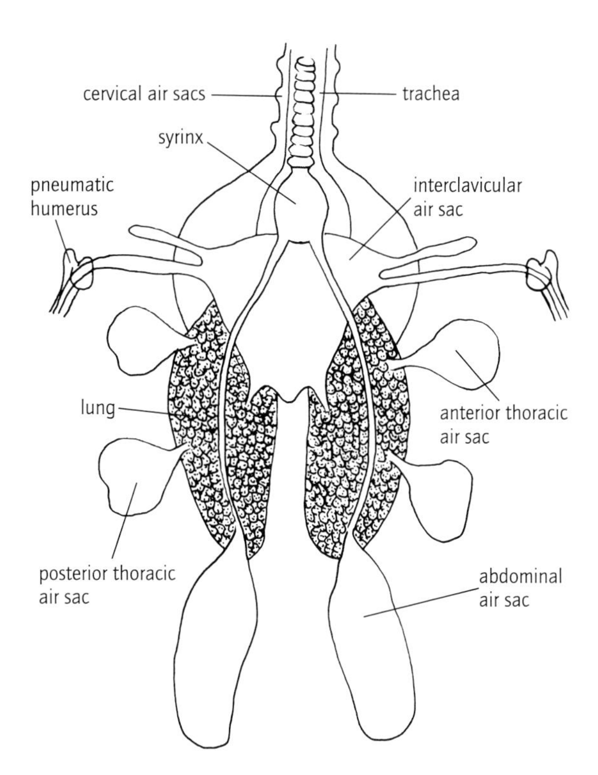


*Plan of waterfowl air-sac system.*

## Lungs and Flight

Compared to mammals, birds have evolved a fundamentally different, and very efficient, system for breathing. Their oxygen demand is high for the energy required in flight, yet Bar-headed geese can migrate over the Himalayas at over 20,000ft (6,000m) where oxygen is thin.

Air enters through the nostrils and trachea, passing down the throat until it reaches the *syrinx* (or voice box). Here the trachea divides into two bronchi which lead the air into the abdominal and posterior air sacs, bypassing the more anterior sacs. When the bird breathes out, the abdomen contracts, and the air moves into the lungs. There follows a further breathing in and out, when air passes from the lungs into the anterior thoracic sacs and interclavicular air sac before exhalation via the bird's bronchi. Thus it takes *two* cycles of expansion and contraction for a parcel of air to pass in and out of a bird's respiratory system.

The air sacs themselves do not function as organs of respiratory exchange: oxygen from the air sacs diffuses into *ventrobronchi* and *dorsobronchi* which connect with *parabronchi.* These in turn connect with air capillaries which are surrounded by a network of tiny blood capillaries, which are responsible for a very rapid gaseous exchange. The avian lung actually allows air to pass through it whilst it behaves almost like a sponge, and the cycle of air flow means that birds can have oxygen-rich air in their lungs all the time.

Unlike mammals, air is not driven into and out of the lungs by means of a diaphragm: in birds, breathing is controlled by muscular contractions of the ribcage which reduce or expand the overall size of the body cavity and thus force air out of the air sacs. So when holding a bird, always allow free movement of the body, because holding it too tightly will affect its breathing.

In addition to the complexity of the air sacs, the hollow avian bones are connected to the air sacs of the respiratory system. Every large bone is pneumatic to a certain degree. This is an important adaptation to reduce the weight of a bird for flight. However, because birds have air sacs that connect with the bones, respiratory infections can spread to the abdominal cavity and bones. It is therefore very important to catch any respiratory infections early in birds, and to avoid high concentrations of ammonia as well as the possibility of fungal infections (*aspergillosis*).

# 5 The Life Cycle and Care of Feathers

## Introduction

We now know that feathers first appeared in non-avian theropods, and so were inherited from these dinosaurs by the birds. They had pliable skin filaments made of the same protein as feathers.

Features once believed to be strictly avian, such as vaned feathers, have also been found. In oviraptors, the short primaries and secondaries clearly had no flight function, and the rectrices of the bony tail were better suited for display than for flight regulation. Vaned feathers and their arrangement as primaries, secondaries and rectrices must therefore have originally evolved for other functions such as temperature regulation, brooding, display or gliding.

Other features once associated just with birds such as down-like body covering, a broad plate-shaped sternum, and substantial enlargement of the forebrain, are also now known to have arisen earlier than true flight. Refinement of the flight capability came rapidly once primitive avians were airborne. Smaller, lighter body size, the evolution of alula feathers on the first digit, a hallux to allow safe perching, and a fused pygostyle at the end of the tail, all confirmed the arrival of flight.

*Structure of a flight feather*

## Types of Feather

### Contour Feathers

Contour feathers cover much of the outer surface of the body. They protect it from physical damage and harmful solar radiation. They act as a 'shell suit' for waterproofing and give efficient streamlining in flight.

### Flight Feathers

Flight feathers are longer and stiffer; they have little or no down. The remiges are found on the wings: the primaries are the feathers of the hand part of the wing, and the secondaries the equivalent of the forearm; the rectrices (rudders) are at the tail.

The central stem of a flight feather is in two parts. The lower part, which originally contained the axial artery as the feather was growing, eventually becomes a hollow tube. This is the calamus, a word meaning 'reed' in Latin. The upper part or rachis has a solid cortex, a centre of pith and a much reduced air space. The underside (ventral surface) of the quill has a groove along its length, contributing to the strength of the structure. The vanes of the flight feathers are asymmetrical; this is most pronounced in the outer primaries.

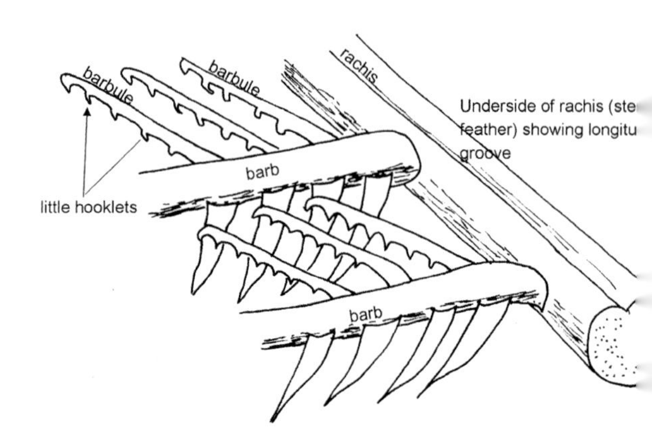


*Barbs and barbules of a flight feather.*

Branching from the stem are the main barbs, which form the basic structure of the vanes. Attached to the barbs are smaller barbules, and attached to these are even smaller barbicels, giving the impression of an interlocking network of branches and twigs all held together with minute 'velcro' hooklets. Instead of tearing like a membrane of skin, the feather fabric simply 'unzips' and is then repaired when the bird runs its bill through the barbs and barbules.

### Downy Feathers

A soft layer of down, beneath the contour feathers, insulates the goose from the cool water and winter cold. Downy feathers have an almost non-existent stem. The long barbs have barbules, but they lack the small hooklets - barbicels - to lock the feather together. They have the ability to expand and trap large amounts of insulating air. Down is thickest on the thorax and abdomen, but covers the rest of the body as well.

### Semiplumes and Filoplumes

In addition there are *semiplumes* for insulation and bulking out the contour feathers, and *filoplumes* – fine, hair-like feathers with a tuft of barbs at the tip. Filoplumes give a sense of touch within the plumage and monitor the position of the pennaceous feathers in the wings and tail. They have sensory corpuscles at their base.

### STRUCTURE OF THE WING FEATHERS

The wing is analogous to the human arm, but the hand region is very much modified by the fusion of bones of the fingers. Twelve secondary feathers are attached to the ulna, and ten primary feathers to the modified 'fingers'. These flight feathers are then followed by the tertials. The flights are covered by greater coverts which are, in turn, followed by the median, lesser and marginal coverts. The marginal coverts are very firmly attached and difficult to remove. When the wing is folded and the bird at rest, the valuable primaries are covered and protected by the tertials.

In the photograph, the tips of these buff primaries have been damaged (faded) by sunlight, whilst the main area of the flights has been well protected: the tertials cover and protect the secondaries when at rest. Buff is not a wild-colour: wild geese tend to be grey. If they are white, they have stronger grey tips to their wings - as, for example, in Snow geese.

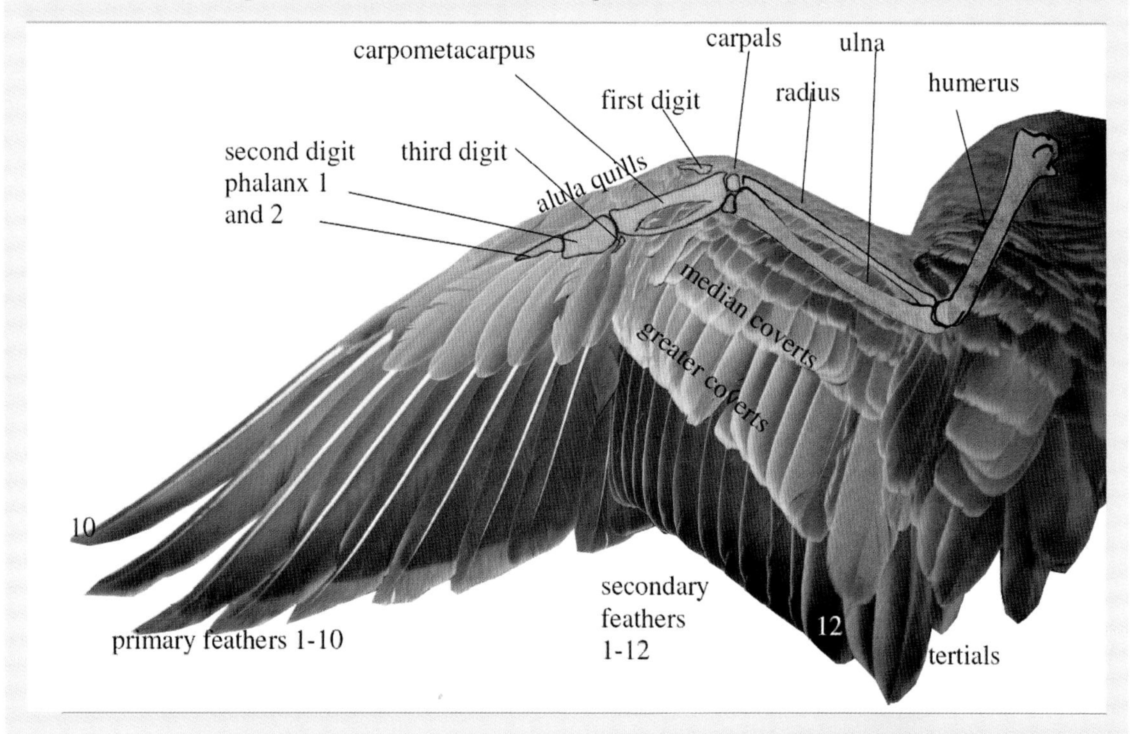

*Downy feathers.*

WHAT FEATHERS ARE MADE OF

Feathers are made of a protein-based material called beta-keratin. The structure is very similar to that of scales, from which feathers may have evolved. Like hair and finger nails, the substance is 'dead'. The material is hard, strong, light, flexible and warm and absorbs very little water.

Sulphur-containing amino acids are important in the structure of the feather protein, and must be synthesized from the birds' diet. That is why a diet containing sulphur-rich amino acids such as methionine is essential for growing new feathers. A methionine-deficient bird will tend to chew or eat feathers in an attempt to satisfy a craving for this amino acid, and this may explain why some goslings behave in this way.

## Moulting

Most Anatids (ducks and geese) have one moulting pattern in common: they all shed their flight feathers on the wings within a few days. This is unlike sequential moulters, which lose just one or two flight feathers at a time so that they can still fly. Geese are flightless when their downy young are flightless too, and they are all very vulnerable on the ground - although their aquatic habitat means that they can escape predation by taking to the water, so the temporary loss of flight is not as crucial as it would be in raptors, for example.

Feathers have a limited life, however much a bird cares for them. But unlike ducks, which moult their contour feathers twice a year, geese only moult their body feathers once. Pair-bonding in geese is very strong so there is no need for bright male plumage to attract a new mate each year; their plumage is not therefore sexually dimorphic, and there is no 'nuptial' plumage, the purpose of which is primarily for male display in species such as the mallard, where the male does not invest in caring for offspring.

The new feathers form from follicles underneath the old feathers. At first there are blood-filled 'pin' feathers, but as they grow the feather pulp retracts to the feather follicle, and the protective sheath splits along its length to reveal the new feather vane, and is removed in the preening process.

Moulting is a very stressful time of year, especially for females who have already lost a great deal of their bodyweight whilst in lay, and whilst incubating eggs. Loss in condition over this period means that their mortality rate is higher than that of males. In domesticated birds it is essential to look after the females very carefully over this vulnerable part of the breeding season.

Non-breeding geese (older non-laying females, and unsuccessful pairs) moult earlier than the breeding pairs. However, the majority of wild birds are synchronized, and timing of the flock's moult is dictated by the breeding pairs.

The sitting goose retains her feathers up to three weeks after the goslings have hatched; she needs to restore her bodyweight before she grows new flights. Then the entire wing tract (primary and secondary feathers) is moulted almost simultaneously, in no particular order, within a few days. The regrowth of these feathers is so important that all other feathers, including the wing coverts, wait until the flights are regrown, in a period spanning three to four weeks. The time of female moult is therefore governed by the date of sitting and date of hatch, so the females are at a similar stage of development to the goslings, which have full flights at seven to eight weeks. In the males, the moulting time is more consistent, and more synchronous, from year to year, and is probably dictated by the hours of daylight.

During the vulnerable time of wing moult, wild geese take to the water, with their goslings, for protection. This timing of the moult is therefore an evolutionary strategy which protects the young. Breast muscle declines, and leg muscle increases as the adult birds become more dependent on running for escape. Goslings, too, focus growth on their legs and gizzard, rather than putting on breast muscle. They have comparatively large feet and strong legs, and can soon run as fast as their parents.

After completion of the new flights, moulting of the body feathers takes place in stages, the broken breast feathers often being replaced first. The last feathers to grow are usually the feathers of the neck, which stick

*As the gosling's new feathers grow, the old feather or down (from the same follicle) is pushed out and then breaks off.*

*Young Brecon Buff gander at about fifteen weeks of age, in August. His flight feathers were complete by eight weeks, and he keeps these until the next year. The first set of body feathers, completed by eleven weeks, was then moulted and replaced by a second set of new feathers by eighteen weeks. His neck feathers have just come through the bottle-brush stage, and he just has to complete the lower neck feathers to attain adult plumage.*

out like a bottle-brush before they finally settle down to their characteristic furrowed, Greylag look.

If individual feathers are lost through mechanical damage, this will stimulate the regrowth of a fresh feather from the same follicle within four to six weeks.

## Caring for Feathers

### Washing and Preening

Wild geese spend a lot of time foraging on land so the bill, legs and feathers can get dirty; however, they always return to their roost on water for safety at night, so there is at least one opportunity to have a cleaning session each day. In a bathing session in shallow water, a goose will submerge its head and fling water over its back. Wing beating on the water's surface is also used to keep the important flight feathers in good condition, and to create a shower of cooling water on a really hot day. In deeper water, geese dive forwards, roll sideways and even overturn. They will characteristically 'yawn', a behaviour pattern which has often been observed but is not understood.

The bathing and cleaning routine is the precursor to the preening process, when the feathers are reconditioned. The outer layer of strong, glossy feathers must be maintained to form a waterproof barrier, thereby protecting the underlying down. Preening therefore includes thoroughly cleaning, inspecting, lubricating and arranging all the feathers at least once a day. The activity includes the removal of parasites, the removal of scales which cover newly sprouting feathers, and the spreading of preen-gland oil. It also involves the combing of the barbs on the feather so that the barbules and even smaller hooks called barbicels all interlock to form a protective water-repellent layer.

After combing the feathers, the preening process ends with fluffing them up well, which may have a dual purpose: to insulate the feather and down structure with air again, and to settle the feathers into the correct place.

### The Preen (Uropygial) Gland

The preen gland, situated on the rump, releases a thick, lubricating oil when the bird stimulates the gland with its bill. The bird then rolls its head and neck over the gland and the oils are transferred to the rest of the body feathers by the stroking action of the bill and neck.

Observations of young waterfowl suggest that preen-gland oil is a water repellent, and young ducklings especially do get themselves and their water very oily. However, feather structure may be more important than the hydrophobic quality of the oil in maintaining waterproofing. During preening, the interlocking feathers are combed so that the barbules and barbicels are ordered at a constant distance from each other. The spaces between them and the adjacent feathers are so narrow that water droplets are not able to penetrate them, and

*Preening involves the removal of parasites such as lice and mites.*

*A two-week-old gosling nibbling the preen gland to release oil.*

it is this constant attention to the zipping of the barbicels of the feathers that makes the waterproof shield.

It has also been suggested that the mechanical action of preening itself applies an electrostatic charge to the oiled feathers. The build-up of charge on the surface of objects happens due to contact and rubbing with other surfaces, and this charge may itself repel the water.

The preen-gland oil does, however, keep the feathers supple. It is also believed that the oil helps kill bacteria and fungi, and there is evidence that it provides a substance which is converted into vitamin D in sunlight on the feathers (Fabricius, 1959). The birds obtain this vitamin by nibbling the feathers during the preening process. Since vitamin D cannot be synthesized on the body of the bird because of the thick covering of feathers, the preen-gland oil replaces this function.

In experiments involving the surgical removal of the preen gland from ducks, lack of preen-gland oil has not necessarily caused waterlogging of the feathers. However, their quality has been affected, and if the feathers become brittle and the barbicels deranged, the water-repellent quality is compromised.

Geese suffer from the condition 'wet feather' less often than ducks. Generally it occurs in older birds and in extreme weather conditions when soft-feathered birds such as Toulouse need to be taken to shelter to dry out. In experiments conducted to find the cause of wet feather, the main reason has always been soiling of the feathers by food, mud, moulds or mineral oils. When a bird dies, the feathers also lose their waterproofing qualities because they are no longer held in place by the bird's muscles, and the air pockets between the feathers collapse, allowing the ingress of water.

# 6 Goose Diet and Digestion

## Introduction

Grass is the basic diet for both wild and domesticated geese. However, the wild birds have a choice of plant species in their diet, and migrate to follow the growing season for the best bite. Thus they not only utilize the high protein first shoots, but also the long daylight hours of the far north to maximize eating time which, if necessary, is almost round the clock. They also secure safer nesting sites, often on islands, and there is security to be found in nesting in a large flock. The benefits of this strategy must outweigh the costs of their long migratory flight.

Domesticated geese, on the other hand, are confined, so it is important that their pasture is well managed and they have supplementary food, especially when the quality of grass is poor in winter, and a greater area of pasture might be needed. Any shortfall in supply must be balanced by grains and pellets. If goslings are being reared, then clean pasture must be set aside for them to thrive over the spring and summer before they are sold and the stock reduced again for the winter season.

## Feeding Habits

Wild geese show a lot of variation in the size and shape of their bill, because they have had to adapt to a variety of foods. The Swan goose has a large, long, heavy bill which is ideal for rooting out tubers and larger leaves, but less well suited to pecking at a tight sward. This is in contrast to the small Red-breasted and Lesser White-fronted, which have short beaks for precision pecking.

The Greylag has an 'all-purpose' bill which can be used for grubbing out roots, grazing pasture or stripping seeds from grasses. They will also paddle their feet on marshy ground to extract roots and stolons of clover. Domestic ganders will do this in the rearing season to attract their goslings.

Geese are inefficient feeders. They are not ruminants and do not make use of the cellulose in the same way

*Red-breasted geese are very small and have not been domesticated. Their short beaks are suited to precision pecking, recorded at up to 180 pecks per minute.*

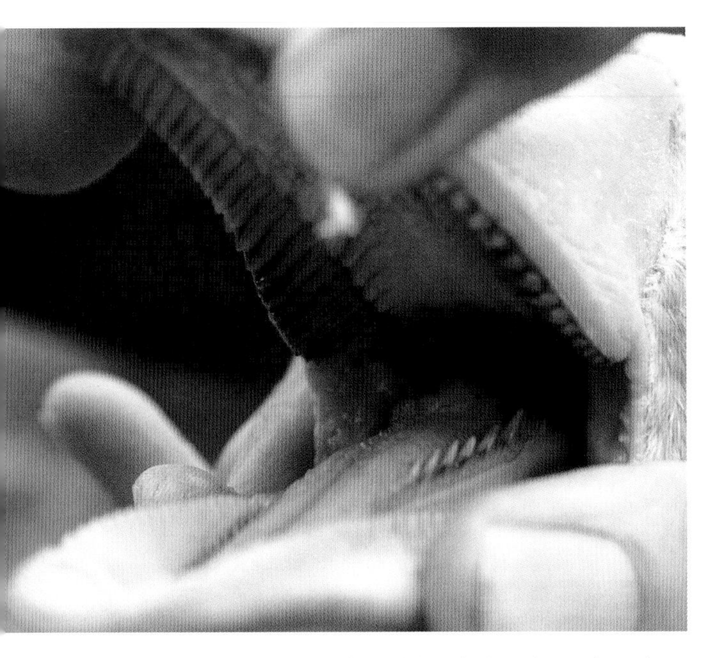

*Lamellae at the side of the bill interlock and are adapted for gripping and tearing vegetation; spikes on the tongue direct the pieces of grass into the throat.*

as cattle and sheep, so they need to gather food over a large part of the day. When daylight foraging hours are short in winter, wild geese must spend seven to eight hours eating. They will graze on moonlit nights, but not in total darkness when their safety would be at risk – they do not have good nocturnal vision. This short foraging time around the winter solstice is the reason why it is essential for the birds to put on as much weight as possible during the autumn in order to reach a peak weight in November. The reserves they build up will help them survive the harsh winter months until the new, high nutrient vegetation growth of spring. So if farm geese are expected to find most of their food from grass, they too will need to spend many hours grazing, and in winter, at least, they will need supplementary food.

## Types of Fodder

Cultivated ryegrass and shoots of winter wheat are staple foods for wild geese foraging on farmland, and they will also eat potatoes and carrots. However, geese will deliberately choose the food that rewards the highest energy return over the winter season. For example, they need a shorter time feeding on cereals in the stubble than on grass to collect their daily food supply. Root eaters also have more time to spare, to rest and watch, rather than eat continually.

As the growing season for grass starts again in the overwintering regions, wild geese first take advantage of this new green bite before leaving for the breeding grounds. As they migrate northwards, in stages, to their nesting sites, they get a sequence of 'first bites' where the spring shoots of reeds, sedges and grasses offer top protein value, at several locations. They are building up nutrient reserves essential for the production of high quality eggs to be laid in the breeding grounds. Note that wild geese lay small clutches of only four to nine eggs, the number depending on the species and the weather and food supply in any particular year.

These seasonal habits of the wild geese, synchronized with food resources, have clear implications for the management and successful breeding of domesticated geese, where diet demand similarly varies over the year.

## Grass Species

On the farm there is mainly grass, rather than a variety of saltmarsh, fen and moorland plants. However, there are varieties of grass and herbs to choose from. In one study of wild geese, grass including some white clover (*Trifolium repens*) is the principal food in spring during the pre-breeding period. My own domestic geese don't particularly like clover, so do monitor the behaviour of your own geese to see what they prefer. Up to 30 per cent clover helps to increase the protein content of the sward as well as improving mineral status by adding nitrogen from nitrate-fixing bacteria.

If grass is to be reseeded, consult a local seed merchant to see what they recommend, because protein content varies according to the species of grass, peaking at 30 per cent in optimum conditions. One of our local merchants recommends a 'mid/late mix' of varieties of rye grass (*Lolium perenne*) with timothy (*Phleum pratense*), generally used for a long-term ley. With cutting and grazing it produces a tight, dense sward and is also very palatable.

Timothy is cited as being palatable but having a lower protein content than other grasses, and rye grass as having 'higher digestibility'. A planting of rye grass alone can result in a certain amount of bare ground around each plant; lower-growing grasses with creeping stems cover the ground better. However, rye grass is successful as long as it is kept closely grazed or mown.

## Grass Quality

Where different qualities of grass are on offer, geese will choose darker-coloured, fertilized, short-length pieces. This is preferred to coarse, low-nutrient grass, and it seems to be recognized by the 'pull factor': thus if a tug at the grass yields a mouthful easily, the grass is desirable and palatable. Even day-old goslings have an inborn preference for green – the brighter the better, while geese feeding in spring will choose the darkest green, fertilized vegetation (Owen 1980, citing J. Kear).

The behaviour of domesticated geese also shows that birds will choose areas of short, clean grass visually as well as by touch. If a new area of sweet grass is made available, the geese will visit, inspect and talk about it before settling down to eat it, confirming its condition and feel by the tug needed to crop it. The tips of short shoots are chosen over longer pieces because they are higher in nutrients - as also in protein and minerals.

Once a good patch of grass is found, the geese will graze it repeatedly while adjacent areas may run to unsuitable long, tough grass. They are mimicking the behaviour of wild geese during the flightless period when they are moulting and at the same time rearing goslings.

Greylag geese have been recorded as re-grazing sward at intervals of between six to eight days, less than the experimentally determined optimum interval of nine days or more, which would give most biomass and protein accumulation. Thus it is clear that geese demand the high protein content of a fresh sward during the time they are also growing new feathers.

Thus from wild goose feeding habits and observation of domesticated geese, the implication is that grass must be kept short to maximize its protein value. They are unlike cattle, which wrap their tongue around longer grass to pull it. So if larger animals are used to graze pasture with geese - and mixed stocking also gives better parasite control - then low biters such as sheep and horses are useful. If the larger animals precede the geese and the grass is allowed to regenerate for one to two weeks, then the goslings get the new shoots. However, don't mix birds with larger stock, which can trample them. Young lambs playing around are especially hazardous for goslings.

Goose droppings from birds on pasture are very frequent, but they only become a problem if the birds congregate at one place. Droppings are just macerated grass, and far less smelly than sheep or cattle droppings. They rapidly disappear in wet weather and do not foul the land as much as is commonly supposed.

Geese cannot be used to get long grass into good condition. Any long grass must be removed first; so control of grass growth by mowing or sheep grazing is essential to keep the shoots in good condition. Thereafter, geese improve grass and they can produce a very short, dense sward.

## Maintaining Grass

It is more important to have good quality grass than a large area of poor, rank grazing. Mowing, weed control, rotation and pH balance are all important.

### Correcting the pH

A pH of around 6 to 6.5 should be maintained for optimal grass growth. If pH measuring equipment is not available, then as a rule of thumb, liming will be useful in the higher rainfall areas of the UK where the bedrock is not limestone. In mid Wales, for example, pasture pH levels are commonly 5.5 or less unless the land is limed. Calcium ions are leached from the soil by rainfall, but also by the application of ammonium nitrate-rich fertilizers and decomposition of plant residues, for example in the goose droppings.

Where pH is very low (acid), greater amounts of agricultural lime will be needed to raise the pH to 6.5. The correct rate can only be applied if the pH is measured using a soil test kit (available from garden centres), and the soil type allowed for. For example, for clay soil at pH 5.5, 350g (11oz) per square metre is recommended for gardens to raise the pH to 6.5. The more acid the soil, the greater the application of lime needed.

www.aglime.org.uk provides a useful application table for larger areas, recommending that grasslands need 5-6 tonnes per hectare for soils of pH 5.5 to raise the pH to neutral status. That works out at 0.5kg (1lb) per square m. The rate is 2-3 tonnes at pH 6.0, and none at pH 6.5. This is for a ryegrass mixture requiring a pH of 5.5-7.0.

Clover will not flourish at pH 5, and its absence is an indicator of low pH. However, an alkaline soil above pH 7 is not an advantage. Too much lime will cause problems with the availability of trace elements. Optimum utilization of soil minerals by grass is around pH 6.5, and a high pH encourages too much clover for the geese.

The Royal Horticultural Society notes that the neutralizing value of agricultural limes such as garden lime and ground chalk (calcium carbonate from pulverized rock) is less than hydrated lime (calcium hydroxide, Ca[OH]2). The hydrated type may also burn growing vegetation, so about half the amount should be used.

Lime is best applied to the garden in the autumn when it can be dug in. But lime has the additional benefit of destroying harmful bacteria so it can be applied to the grass surface prior to gosling grazing in spring, or to help clean pasture which has become dirty and is being rested. In a small area, apply it as a fine dusting, spread from the garden shovel. Take care to apply the fine, dusty lime on a day when wind speeds are low, and wear protective glasses and gloves because the dust (especially hydrated lime) is an irritant. It must be well watered in and washed off by rain and new shoots grown before grazing again.

A longer-term solution is to apply calcified seaweed for a slower release of nutrients; the geese will also enjoy nibbling the particles if they need the nutrients.

Extraction of this material may, however, be environmentally damaging, and it is best used in small amounts as a supplement with the mixed grit (*see* Chapter 16).

### Controlling Weeds

There are certain plants that geese cannot eat; for example they avoid docks, nettles, thistles, mayweed and daisies, and generally do not eat plantain. Mowing not only controls weeds, topping them before they develop seed boxes, but also keeps grass finer and encourages new, protein-rich shoots.

Selective weedkiller may be needed on very weedy ground but must be used carefully: if stock is turned out too soon, the application will blister their mouth and feet. Directions for the use of herbicides generally advise keeping stock off the ground for a minimum of ten days following treatment, and in fact it is best to vacate the ground for much longer. After leaving the weedkiller to take effect, cautious goose breeders will mow and discard the grass cuttings from the following month's growth because the weedkiller does not break down rapidly: it is taken up by the plants and therefore ingested by the grazers. Although not supposed to be harmful to animals, it does stay in the food chain for some time.

For example, Clopyralid, used as a selective herbicide for the control of broadleaf weeds, is known to persist in dead plants and compost. Triclopyr is also used to control broadleaf weeds while leaving grasses and conifers unaffected. It breaks down in soil with a half-life of thirty to ninety days; however, by-products remain in the soil for up to a year. It is therefore prudent to discard mowings in a place not used for animal grazing or for garden composting. Herbicides are best used as a last resort, and certainly should not be used on an annual basis.

## Food Digestion

The digestive system begins at the bill and ends at the vent, and also involves the liver, pancreas and caeca. The bill tears up the grass - the basic food - which is then swallowed and passed down the oesophagus.

Unlike the chicken, the goose does not have a distinct crop to store food, but the oesophagus and the proventriculus (where the alimentary canal broadens before it joins the gizzard) fulfil this function. In the evening, geese will often stuff this long tube from mouth to gizzard before they are shut up, so they have an overnight food supply; some birds will eat so much that their neck can look deformed and feels absolutely solid. This can temporarily impede their breathing. They do this when young because goslings have such a huge food demand in order to keep up their growth rate. Adults also behave in the same way in the winter months when nights are long. The gut has a very rapid throughput of only two hours, and they are attempting to make their food supply last as long as possible. Wild geese also behave like this, collecting and storing food before returning to their safe roost for the night

Passage of the food down the oesophagus is aided by mucous glands to lubricate it, and gastric secretions (hydrochloric acid and the protein-digesting enzyme pepsin) to aid the digestion process in the proventriculus. Food then moves into the gizzard for further preparation, but can move backwards and forwards between these two organs to optimize digestibility of the particles.

### The Gizzard

It is particularly important that grazing geese have access to hard grinding material for the gizzard. The gizzard is a giant muscle used to mill the food and make it suitable for the intestine to extract the nutrients. Since the beak only snips off bits of grass, the gizzard, in the absence of teeth, has to comminute the material into small particles so that nutrition can be more easily absorbed in the intestine. The goose does not have the complex digestive system of ruminants, so the gizzard breaks down the cell walls of the grass by physical action so that the juices are released.

To extract the maximum amount of nutrition, the large, muscular organ of the gizzard exerts very high pressure on the food, enabling the goose to make much better use of fibrous food than other farm species such as the duck and hen.

Wild geese spend time finding suitable material for the gizzard, and goslings just a few days old will sieve sediment in puddles to find the right grade of material. Mixed poultry grit is therefore essential to act as milling material or 'teeth', but for the geese, coarse sand is equally important. In an area of sandy soil, the birds will eat this, or help themselves from piles of builders' sand.

Both grit and coarse sand must be available all year; if geese do not have access to grinding material, they will not thrive. Characteristically, about 15g (¾oz) of sand and stones, together with fibrous debris, are found in the gizzard (Mattocks 1971). If geese have been without grinding material for some time, they will devour large quantities of sand very quickly, and it is probably better to ration it in these circumstances.

Birds seem to suffer less from gizzard worm (*see* Chapter 20) if they have access to proper grinding materials. The gizzard lining is quite coarse and horny due to the action of grit, and this layer sloughs away and re-forms periodically. This may well keep the gizzard healthy and is a process that will be helped by regular access to suitable materials.

## The Intestine

Milled food or 'digesta' is moved on to the small intestine (duodenum), where further digestion, of the cell sap released by the puncturing of grass in the gizzard, takes place. The pancreas secretes protein-digesting enzymes so that nutrients from the food can be absorbed through the walls of the gut and into the bloodstream, to be used for growth, repair and energy.

At the end of the small intestine are the caeca which are well developed in the goose that digests large amounts of fibrous food. They contain bacteria that were thought to break down plant cellulose into digestible carbohydrate. The caeca are followed by the large intestine, which leads to the cloaca where faecal material is eventually ejected through the vent.

The food input to the system is so great that geese (measured in wild Barnacles and White-fronted) excrete a dropping every 3½ minutes over the eating period (Owen 1980). Where the calorific value of food is higher – as, for example, in cereals – droppings are fewer.

It was once assumed that the goose had a digesting mechanism enabling it to use cellulose as an energy source. In ungulates such as cattle and sheep, food is allowed to ferment in the rumen so that the cellulose can be converted by bacteria and ciliates into appropriate nutrients. By contrast, the rapid throughput of the goose (two hours compared with twenty-four to forty-eight hours for the whole passage in the ruminant) gives little time for such complex digestion.

Mattocks (1971) investigated the alimentary canal of the domestic goose to see if bacteria played any part in digestion of the fibre of the grass. Previous authors had thought that the extension from the large intestine known as the caecum might functionally resemble the rumen. He extracted material from goose caeca and used it to try to digest plant fibre. This was not successful, unlike the extract from the rumen of a cow where, using the same technique, cellulose fermentation took place.

He came to the conclusion that the caecum was a 'congenial place for anaerobic bacteria' but showed no evidence of using food, and with a capacity of 20ml, could not handle the large volume of food required daily.

Subsequent research investigated the microbial breakdown of hemicellulose in the large intestine and in the caeca. While cellulose is crystalline, strong and resistant, hemicelluloses – which are embedded in the cell walls of plants – are broken down by dilute acids and enzymes. Waterfowl can metabolize 25–74 per cent of hemicellulose in plants. It was estimated that Pink-footed geese were able to digest 24–29 per cent of dry matter in grasses, and 53–64 per cent in barley grains (Krapu and Reinecke, 1992).

## Supplementary Food

Grass should form the basic goose diet, and it is especially important for the start of the breeding season, and for rearing goslings. However, domesticated geese must be fed additional rations to stay healthy. A typical wheat/pellet mix of 200g per day may be needed for geese in winter, depending on grass quality (*see* Chapter 16).

The stocking density for geese on pasture will vary

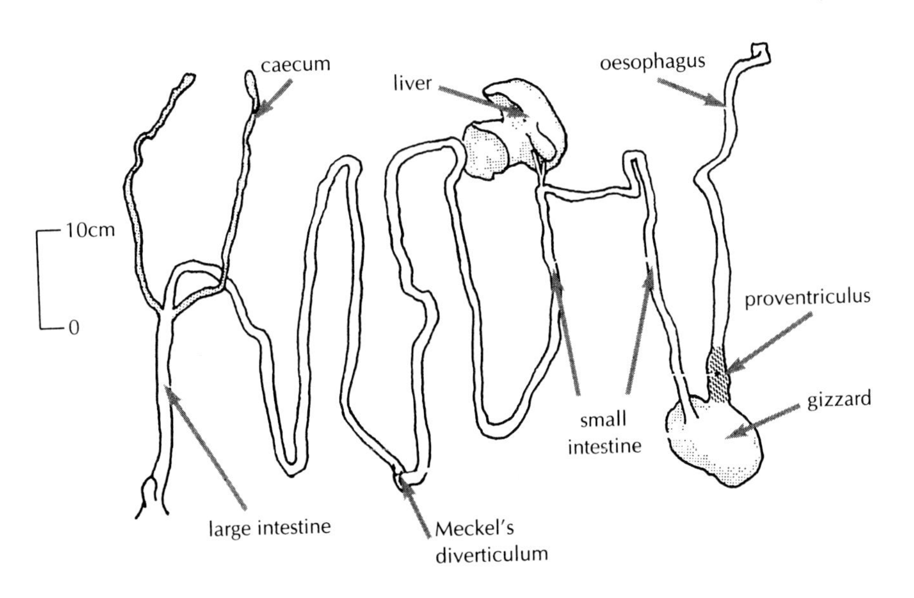


*The gut of the domestic goose. Food is stored in the oesophagus and proventriculus before passing into the gizzard for milling (Mattocks, 1971).*

depending on the quality of the pasture, the time of year and grass growth, the amount of supplementary food, and the age and size of the geese. Figures vary widely between 125 to fifty birds/ha for growing geese, and twenty birds/ha for breeding geese. The lower figure is preferable at all times because the grass stays cleaner, and this will avoid coccidiosis.

KEY POINTS

* Geese raised on grass alone must have several hours a day to obtain their nutrition.
* Wild geese have a varied diet and select different plant species according to seasonal nutrient supply.
* Autumn and winter food can also include higher energy root vegetables and stubbles.
* Young shoots of grass provide more protein than longer grass.
* Grass quality should be maintained by cutting and liming.
* Selective herbicides to control weeds should be used with great care.
* The gizzard is especially adapted to grind up grass.
* Geese and goslings need grit for the gizzard to release cell sap.
* Domesticated geese also need supplementary food in their diet.

# Part IV
# The Breeds of Geese

## Introduction

Domesticated 'geese of the land', known as 'landrace', have been found over a greater part of Europe and Asia since time immemorial. The type kept by the local people depended upon the wild species they used, but also upon the characteristics that people chose, and upon the effect of the local environment. In China, for example, the egg-laying strain of the Chinese was developed, as well as the larger strain of the 'African' for the table.

In Europe, white geese and curly-feathered geese were traditionally selected, and the saddleback and auto-sexing types were also developed. These forms of domesticated geese were chosen from mutations that were noticed in the domesticated breeding flock - selection was not based on any knowledge of genetics, but upon cultural preference and tradition. Thus white goose down was preferred in northern Europe, for duvets; but a white gander and a grey goose were the general maxim in the west, for ease of auto-sexing.

The concept of the 'landrace' is thus well established in Europe: it is literally the 'goose of the land', meaning the typical bird of a certain district - the bird which is best adapted to the local conditions of climate, management practices and food supply. Where these breeding populations were extensive, a wide gene pool and inevitable cross-breeding maintained productivity.

As will be seen in the following chapters, breeds of geese are now divided into descendants of the Swan goose, comprising the Asiatic geese - the African and the Chinese (as described in Chapter 8) - and descendants of the Greylag goose, including Several white breeds, the Sebastopol, autosexing and pied breeds, the Toulouse, buff breeds and fighting geese (as described in Chapters 9 to 15).

# 7 Introduction to the Breeds

## Native Breeds

Quite often it has been from the landrace that a breed has been developed and refined for the standards book – as, for example, the German Embden and French Toulouse. Such developments are known as *native breeds* – that is, they have a particular place of origin. Today there are guidelines on the definition of a native breed for larger animals such as cattle and sheep; these include the following (simplified) points for the Rare Breeds Survival Trust for the UK:

* The breed must comply with the FAO and EU definitions of a breed.
* Breed history documents the breed origin and development within the UK.
* Breed history documents its presence in the UK for forty years plus six generations.
* Not more than 20 per cent of the genetic contributions come from animals born outside the UK (other than those imported for an approved conservation project) in any generation for the last forty years plus six generations.

Some native breeds came about from improvement in breeding stock during the agricultural revolution. Changes in land use and productivity, and redistribution of the human population, altered both products and markets. Commercialization for the urban market must have refined the processes whereby a group of animals was developed as a uniform, mass-produced, profitable product. That is well documented in the case of the Aylesbury duck.

Today, native breeds may no longer be used as commercial birds, nor be as genetically diverse as the original landrace – they may even *be* rare breeds because they have been neglected commercially. Some of our native goose breeds have only recently been recognized. The West of England, Shetland and Normandy goose are good examples of native breeds which fell out of favour, and have now perhaps been rescued due to their rare breed status and eventual standardization.

In Germany, where the goose culture was strong, there were many regional types, documented by Bruno Dürigen and Horst Schmidt. Well fed, heavier geese such as the Embden were preferred on the heavier soils of the North Sea, and the Pommern goose on the Baltic. Elsewhere were many regional types, such as the Diepholz, Leine and Lippe geese. Some types came in a variety of colours – white, grey, buff, blue and pied.

## Rare Breeds in the UK

Not a great deal of significance has been attached to breeds of geese in the UK, probably because they are of no great economic importance. However, the incidence of notifiable diseases such as avian influenza and Newcastle disease in the last ten years has led to a formal definition of a rare breed by Defra. In theory, this status should protect rare breed stock from automatic culling during an epidemic. Rare breeds must:

* be a native domestic breed predominantly or significantly domiciled in the UK. The term 'significantly domiciled' has been included to cover breeds where the UK holds the world's largest viable population
* have fewer than 1,000 pure-bred females in the UK
* have bred true for a significant number of generations (fifty years)
* be recognized by a governing body and accepted as a standardized breed

Rare breeds in 2021 include the auto-sexing Pilgrim, Shetland, West of England; also Brecon Buff, Buff Back and Grey Back, Embden, Roman, Sebastopol, Steinbacher and Toulouse.

## Breed Standards

A breed is a group that has been selected by humans to possess a set of inherited characteristics that distinguish it from other geese (*see* Chapter 3). The formal idea of a 'breed', and thus a *written 'breed standard'* for birds, is a comparatively recent concept; it probably followed the idea of the 'landrace' from which some formalized breeds, such as the Toulouse, originated. Note that selection within a landrace for the consistent

*Germany has more breeds of geese than any other Western country because of its long tradition of 'geese of the land', and also its size of territory. Some of the breeds are a type represented in different colours, as, for example, Alsace (Elsässer) geese in the traditional grey (a), and now pied (b).*

appearance of a breed narrows the gene pool of the stock, and the characteristic productivity of the landrace can be lost.

Standards could only arise in a society connected by travel and trade with other regions, and a realization that a type of animal commonly found in one area could be significantly different from those found elsewhere. International travel and the import of exotic birds made exhibition, competition and formal standards desirable and possible in Victorian times, when indeed the first standards for poultry and waterfowl were published in the UK and USA. The central purpose of defining a breed in this way was to set guidelines for competition and judging at exhibitions (*see* Chapter 2).

The description of a breed may vary from country to country, both in format and content. The format of the USA standards has been consistent over time, whereas that of the UK has been idiosyncratic. The rather eclectic set of earlier UK waterfowl standards has only recently been rationalized into a common format, a process which began in 1997 and was finalized in 2008.

For a particular breed, the basic description will nevertheless not vary a great deal from one location to another, or over time. For example, the description of the Shitou in China is very similar to the African in the USA. Furthermore, the African standard in the UK was adopted from the USA. Thus there is a useful tendency for standards, wherever they are, to adopt an agreed norm. This is generally based upon the description of the country of origin.

Standards give a short history of a breed, including the origin. They describe the outward appearance - the phenotype - in shape, carriage, colour and feather type. Although weight and size are given, there is generally no reference to productivity in egg production, hatchability or growth rate. This is in contrast to commercial flocks where productivity - for instance in egg

| *Goose name and country of origin* | *Standardized breeds of geese* | *Chapter reference* |
|---|---|---|
| **Alsace**<br>*Border of Germany and France* | *France:* Alsace<br>*Germany:* Elsässer - grey and pied | |
| **African - Shitou**<br>*China* | *Australia:* African<br>*France:* L'Oie de Guinée<br>*Germany:* Afrikanische Hockergänse<br>*UK:* African - Brown 1982, Buff 1999, White 1982<br>*USA:* African - Brown 1874, Buff, White 1987 | 8 |
| **American Buff**<br>*USA* | *UK:* American Buff, 1982<br>*USA:* American Buff, 1947 | 13 |
| **Auto-sexing**<br>*Europe* | *UK:* Pilgrim 1982<br>*USA:* Pilgrim 1939<br>*Australia:* Settler 2010<br><br>*France:* Normande (Sequanian), Bavent (Huppé)<br>*UK:* West of England 1999 and Shetland 2008<br>*USA:* Cotton Patch and Shetland | 11 |
| **Bohemian** | *Germany:* The 'Czech' goose developed in Germany has now been renamed as Bohemian<br>*UK:* ditto | 9 |
| **Bourbonnais** *France* | *France:* Bourbonnais | 9 |
| **Brecon Buff** *Wales, UK* | *UK,* 1954; 1934 *The Feathered World* | 13 |
| **Bresse** *France* | *France:* Bresse | |
| **Chinese**<br>*China* | *Australia:* Chinese<br>*France:* L'Oie Caronculée de Chine<br>*Germany:* Hockergänse<br>*UK:* Chinese 1922, 1954<br>*USA:* Chinese, 1874, Brown, White | 8 |
| **Celler** *Germany* | *Germany:* Celler Gänse | 13 |
| **Czech** *Czech Republic* | *Germany:* Tschechische Gänse | 9 |
| **Deutsche Legegänse** *Germany* | *Germany:* Legegänse (egg-laying goose) | 9 |
| **Diepholz** *Germany* | Diepholzer Gänse | 9 |
| **Embden**<br>*Germany* | *Australia:* Embden<br>*Germany:* Embdener Gänse<br>*UK:* Embden 1865<br>*USA:* Embden 1874 | 9 |
| **Emporda** *Spain* | *Germany:* Emporda-Gänse | 9 |
| **Franconian** *Germany* | *Germany:* Fränkische Landgans - blue (also white, pied) | 3 |
| **Landes** *France* | *France:* Landes | |
| **Lippegänse** *Germany* | *Germany:* Lippegänse | |
| **Pied**<br>*Europe* | *France and Belgium:* Oie Flamande, also in grey<br>*Holland:* Twente landrace (1895), also in white<br>*Sweden:* Oland, Skånegås<br>*UK:* Grey Back (Pomeranian)/Buff Back 1982 | 12 |
| **Poitou**<br>*France* | *France:* Poitou | 9 |

| | | |
|---|---|---|
| **Pommerngänse** *Germany and Poland* | *Germany:* Pommerngänse - grey 1912, pied, white<br>*UK:* Pomeranian - pied grey 1982, grey, white 2008<br>*USA:* Saddleback Pomeranian in grey and buff, 1977 | 12 |
| **Roman** *Europe* | *Australia:* Roman<br>*UK:* Roman 1954; crested Roman 1999<br>*USA:* Roman 1977, tufted | 9 |
| **Russische Gänse** *Germany* | *Germany:* Russische Gänse<br>(a grey goose, not the Tula or Arsamas) | |
| **Sebastopol** *East Europe and Russia* | *Australia:* Sebastopol<br>*France:* L'Oie Frissé du Danube<br>*Germany:* Lockengänse<br>*UK:* Sebastopol - White 1982, Buff 1997<br>*USA:* White 1938 | 10 |
| **Steinbacher** *Germany* | *Germany:* Steinbacher Kamfgänse, Grau, Blau<br>*UK:* Steinbacher - blue 1997, grey 2008<br>*USA:* Steinbacher | 15 |
| **Toulouse** *France* | *Australia:* Toulouse<br>*France:* Toulouse<br>*Germany:* Toulouser Gänse<br>*UK:* Toulouse - Grey 1865, White 1982, Buff 1997<br>*USA:* Toulouse - Grey 1874, Buff 1977 | 14 |
| **Touraine** *France* | *France:* Touraine | 9 |
| **Key:** | Popular, distinctive breeds, worldwide standards | |
| | Buff breeds | |

The commercial breeds of Russia and China, listed in FAO documents, are not included in this table. The major interest in these countries is in commercial flocks where cross-breeding is widely used to maintain heterosis.

The lighter-weight geese from China such as the Huoyane (3.7–4.5kg/8–10lb for males and 3.5–4.3kg/7.7–9.5lb for females) average 100 eggs per year. In contrast, the larger table breeds, such as the Lion Head or Shitou, weigh 8.85kg (19.5lb) for males and 7.86kg (17.3lb) for females, and produce an average annual egg number of twenty-four per goose (FAO report on AGR in China, 2003)

In the FAO report on Russia (1989), some pure breeds are cited - for example the Sebastopol and Toulouse - but the commercial farms are more likely to raise geese produced from recent crosses with Chinese, such as the Kuban. It is also stated that most farms 'do not raise purebred Chinese geese but their crossbreds with other breeds which have higher meat qualities.' Note that, although the USSR FAO report does refer to 'breeds' such as the Kuban, other types are referred to as 'breed groups'. In contrast, in China, all twenty-six groups described are referred to as 'native breeds'.

Goose-producing countries such as France and Germany also have regional non-standardized landrace geese, though these may not be particularly distinctive out of their regional context. Schmidt (1989) lists the Wetterauer, Kurhessische, Main, Ulmer, Oberbayerische, Probsteier, Elb, Sachsenland, Oberschlesische, Leine, Lippe and Niederrheinische Legegans as other types of German agricultural geese. Regional types were also listed in Dürigen (1923).

Note that some breeds are recognized internationally. They tend to be those which are distinctive in size, type and colour. In the white geese, it is the large Embden and the small Czech (now Bohemian) which are standardized. Also included are the 'pure' Asiatics. However, where their crosses are particularly beautiful and distinctive (as in the Steinbacher) there has also been international recognition.

Buff breeds are also desirable, and have produced three different breeds.

production and in food conversion - is measured. The physical phenotype of the commercial strain is of less importance than its productivity.

In the standard, a list of commonly observed faults is often given, such faults being indicative of cross-breeding - that is, with another breed. Any deformity and the worst faults will disqualify a bird.

At present in Europe the *European Entente* is trying to establish uniform standards across the EU, based upon observations in the country of origin. Nevertheless, there are still differences in breeds such as the Emden (German and English style) and Toulouse (French, German and UK/USA style).

As well as documenting traditional types of bird, standards also admit new breeds from time to time. Some of these are imported, and some are old breeds which have been neglected and almost lost - for example the Shetland.

Some 'new breeds', such as the Brecon Buff in 1934 and the Steinbacher in 1932, have arisen from 'the geese of the land'. Whenever a new breed is created, it must be phenotypically distinctive, be desirable, and breed true. These criteria set standardized breeds apart from commercial strains and 'landrace' geese. However, some of the old 'landrace' types enter the standards if they are distinctive and become popular for exhibition, as, for example, the Blue Franconian.

## Goose Plumage: Texture and Colour

Originally, many of the breeds were partly defined by their colour. Thus an understanding of goose colour genetics is useful, both in defining the standards for the birds, and in breeding them. Birds that appear to be true representatives of the breed should breed true, but certain genes are recessive and may re-emerge in subsequent generations. Breeders of pure lines of exhibition breeds are aware of these tendencies through years of practical experience and record-keeping. Note that the genetic characteristics of the colours and patterns of goose plumage have been assessed by observing what is bred, from a controlled gene pool, through several generations.

### Colour Mutations

DNA analysis confirms that European domesticated geese were derived from the wild Greylag. Western Greylags have a more orange bill than the eastern types (*see* Chapter 1), thus giving rise to both pink and orange-billed breeds.

The plumage pattern of the Greylag is the basic colour form, as in the Toulouse, and is referred to as 'wild-colour'. However, white geese have been selected from early times. The 'white' may have been similar to the Embden and Roman, where both males and females are white. On the other hand it could have been auto-sexing white where males are white and females are grey, or partly grey.

The colour and pattern of Greylag goose plumage is genetically simpler than in ducks. There is grey, white, auto-sexing white and pied. Grey is the basic wild-colour; buff and blue dilutions are additional to these.

Certain colour mutations are typical of Chinese and African types, i.e. Swan geese. The blue mutation may have been derived from these birds.

The inheritance of colour in geese was described by Robinson (1924), Gordon (1938) and Oscar Grow (1961). It was investigated further by Jerome (1954, 1970), and his work was summarized by R. G. Hawes (1990). A summary of this work is given on the table below. Examples of the colours, and how they breed, are given in the appropriate breed chapters.

| *Key* | *Colour Mutations in Domesticated Geese* |
|---|---|
| ($Sp^+$) and ($G^+$)<br>($sp$) and ($g$): | $^+$ **denotes wild-colour** i.e. *solid pattern, grey*<br>**Capitals denote dominance**<br>**Lower case denotes recessive e.g.** *spot* and *buff* mutations |
| | **Mutations in Greylag Geese - *Anser anser*** |
| Solid Pattern ($Sp^+$) | This term denotes solid-coloured birds such as the ***Toulouse.*** The original grey colour is also called *wild-colour*. The **Brecon** and **American Buff** are also ($Sp^+$) plus the buff gene. |
| Recessive Spotting ($sp$) (allelic to the above) | This was referred to as sex-linked; that is now known to be incorrect. Both males and females ($sp/sp$) have areas of wild-colour on the heads, upper necks, scapular features and thigh coverts. The remaining plumage is white. |
| Sex-linked dominant dilution ($Sd$) | In combination with ($sp$), ($Sd$) produces white geese such as the *Embden, Roman* and *Czech*. The white of the Chinese is different: it is a single recessive gene. White European geese are $Sd/Sd$, $sp/sp$ in males; $Sd/-$, $sp/sp$ in females.<br>Where the spot gene is absent, solid pattern males $Sp^+/Sp^+$ with dilution $Sd/Sd$ appear almost white i.e. ***Pilgrim male***. In the ***Pilgrim female*** $Sd/-$ the single dilution gene just reduces the grey wild-colour to lighter grey. |
| Sex-linked modified dilution ($Sd\#$) | In auto-sexing spot pattern breeds such as the Normandy, Shetland and West of England (WoE), sex-linked dilution is modified. Dilution is less effective, denoted ($Sd\#$). It fails to whiten areas not affected by white spotting in hemizygous females, and results in dilute grey feathers in the pied pattern. Homozygous ($Sd\#$) feathered males are white. |
| Sex-linked recessive buff ($g$) | This is allelic to wild-type grey in European breeds (Jerome, 1954). The feather pattern (paler-edged) remains the same. The buff gene is found in Brecon and American Buffs ($Sp^+/Sp^+$, $sd/sd$ - no spot and no dilution). The buff gene has also been passed on to *Toulouse* and the German *Celler*, also the *African* (where the general feather pattern is also Asiatic).<br>In the *Buff Back* and *Buff Pomeranian* the genotype is ($sp/sp$, $sd^+/sd^+$) with the sex-linked buff gene (g). |
| Blue (Bl) | Blue was perhaps first recognized in the German Steinbacher. The colour is autosomal and incompletely dominant. A grey crossed with a homozygous blue will produce 100 per cent offspring in the darker heterozygous blue. Different shades of the heterozygotes indicate that modifier genes vary the expression of the colour. |
| Sebastopol plumage (Sb) | Elongation, fluting and curling of the feathers is caused by the Sebastopol gene (*Sb*). The rachis of the feather is also affected. Birds homozygous for the gene show the full curled expression. The gene is autosomal and incompletely dominant.<br>(M. Ashton, 2009) |
| Tufted/crested | Crested geese can arise in any variety, but are most frequently seen in the Roman in the USA and the Emporda in Europe. As with crested ducks, the cranium is sometimes adversely affected. |
| | **Mutations in Swan Geese - *Anser cynoides*** |
| Autosomal recessive white (c) | Pure birds (c/c) are white; gosling down is yellow. Eyes are blue, bill and legs orange. Heterozygotes show coloured feathers. |
| Autosomal neck stripe (Ns) | This feature is seen as a dark brown stripe in the Swan goose, African and Chinese. Cross-breeds with greylags exhibit this feature, albeit less intense. A faint neck stripe is retained in the Steinbacher. |
| Breast patch *(Wb)* | This does not appear on pure breeds, but is expressed in F1 hybrids of *Anser cygnoides* and *Anser anser*. It was thought to originate from *Anser cynoides*, and said to be absent from Pilgrim × Roman (*A. anser*) though such crosses in the UK have produced (*Wb*). |

*Based on Hawes and Jerome*

# 8 Asiatic Geese

## Introduction

For many centuries, contact between the Far East and Europe was limited to valuable products transported overland. These were unlikely to have included live birds or eggs, but with the opening up of sea routes to the Far East, the exchange of bulkier and lower value commodities became possible, and birds could have been kept alive by interested keepers, as with the Indian Runner duck, aboard ship.

Asiatic geese include the Chinese and the African goose, both of which were derived from the Swan goose (*see* Chapter 1); both had reached Western Europe by the 1600s. A goose very much like a Chinese cross-breed is shown in a d'Hondecoeter painting of the 1600s. Also, in Ray's English translation of *The Ornithology of Francis Willughby* (1678), the 'Swan goose (*Anser cygnoides*)' is described, but not illustrated.

Many of the early names, descriptions and illustrations described birds of mixed colours (white crossed with the wild-colour of the Swan goose), various crosses with European geese, birds of varying sizes (between the larger 'African' and the smaller Chinese), and birds with and without dewlaps. Dewlaps were sometimes referred to as 'wattles', and a variety of names were used for the birds. This variety of ideas is quite understandable, because as well as mixing the large and smaller types and colours, the specimens arrived by different routes – both overland from Russia and by sea from the Far East.

The French naturalist Buffon (1793 translation) continued to confuse the situation by referring to 'l'oie de Guinée' (*Anser cygnoides*), a rare swan-like goose from exotic parts. Dixon (1848) also states: 'The old writers call it the Guinea goose, for the excellent reason, as Willughby hints, that in his time it was the fashion to apply the epithet "Guinea" to everything of foreign and uncertain origin'. The name of 'Guinea' has been translated as 'African', and this has stuck ever since.

As late as 1835 Jenyns still referred to *Cygnus Guineensis* (Guinea swan) but also to *Anser Guineensis*, Chinese goose and Swan goose: 'Said by some to be originally from Guinea. Of restless habits and very clamorous.'

Dixon, who wrote from first-hand knowledge, concluded: 'A goose, however, it decidedly is, as is clear from its terrestrial habits, its powerful bill, its thorny tongue, and its diet of grass. And therefore we have determined to call it the China goose concluding that Cuvier is right about its home, and other authors about its goosehood.' Cuvier had found that: 'In journeying overland (in books of Travels) we met with the Swan Goose more frequently as we approach Tartary and China' (Dixon and Kerr, 1853).

By the mid-1800s the term 'Guinea' seems finally to have been dropped, and distinct varieties and colours were then described. In 1848 Dixon was told by the head keeper of the London Zoological Society that there were three varieties, namely the White and two sorts of Browns (Edward Brown, 1906). This evidence of three kinds was substantiated by Nolan (1850), who also correctly ascribed all of the 'knob geese' to Asia; he distinguished the large Brown Hong Kong goose (African), and both White and Brown Chinese.

## The African

Despite the illustrations in Nolan and Dixon and Kerr (below), some authors persisted with the idea that the 'African' (Nolan's Hong Kong goose) was produced by cross-breeding the smaller Chinese goose, for its colour, with the larger Toulouse for its dewlap. However, as Grow (1972) points out, there is no need to invoke the Toulouse to produce the African, since African specimens with good dewlaps pre-date the evolution of the decided dewlap on the Toulouse.

It is also likely that African geese reached the USA before the Toulouse were imported. A letter from Mr Belcher, dated 1852, is quoted in Robinson (1924):

*OPPOSITE PAGE: Seventeenth-century Chinese goose with Muscovy pair and duckling. Chinese geese were in Europe and possibly the UK; Muscovys had been imported from South America. Melchior d'Hondecoeter, by permission of Burton Agnes Hall, Driffield: www.burtonagnes.com*

*In 1734, Albin illustrated the 'Moscovian', having acquired it from a 'Moscovian Merchant, who sent for them to that Country with a Design to propagate them here, which he did and sold them at a great price'. The prints show a male with a dewlap (like the African or Shitou), and a female more like a Chinese. Both birds appear to be brown/white crosses; note that the pied male is similar to the pied Russian Kholmogorsk. The pictures of the Moscovians were probably drawn from specimens brought to Albin, the inference being that the birds were in England at the time.*

> The China geese, which I have bred for some time past, are generally considered natives of Hong Kong, and are often named after that place; but those who have been in that region, and have had inducements to observe the fact, concur in stating that there is no ground whatever for the assumption that Hong Kong is their native place . . . The breed that I own, which possess great merit, were brought from Tchin Tchu and have therefore the best right to the possession of the euphonious title associating them with their place of origin . . . As respect to their properties, they grow to the weight of from forty or fifty pounds a pair at mature size – say at two years old.

Mr Belcher indicates that his geese were large and must indeed have been the 'African' (not 'Chinese'). They came from the province of Fuhkien, opposite Formosa (Taiwan).

The Chinese and African were both standardized in the USA in 1874 (APA Standard of Perfection). Strangely, the weight of adult Africans has been kept at 20lb (9kg) for the gander and 18lb (8kg) for the goose, despite the larger weights for the breed in China and in the show pen both in the USA and the UK.

## The Tse Tau or Shitou

It seems obvious from the colour of the birds that the African is a relative of the Chinese. Both are coloured as the wild Swan goose, and the Chinese behave and call in a similar manner.

An examination of the chromosomes of the birds was one obvious way to prove that the African actually originated in China. Hoffman (1991) cited the work of Bhatnagar (1968) to prove the consanguinity of the wild Swan goose and the Chinese goose. The work of Silversides also proved that the African is also the descendant of the Swan goose: 'All three manifest a unique inversion of the fourth autosome not present in the karyotype of domestic geese descendant from the wild Grey Lag':

> Hybrids were produced between an African male and several Pilgrim female domestic geese. Partial *karyotypes revealed a difference in the fourth largest pair of autosomal chromosomes. This chromosome pair was metacentric in the African, submetacentric in the Pilgrim, and heteromorphic in the hybrids . . . These findings provide cytological evidence to support the traditional opinion that the African breed was derived from the Asiatic swan goose (*Anser cygnoides*) and the Pilgrim breed was derived from the European greylag goose (*Anser anser*).
>
> (Silversides et al. 1988)

[*karyotype: appearance, under a microscope, of chromosomes in the nucleus of a eukaryote cell. The appearance of the chromosomes can be used to study taxonomic relationships, and to gather information about past evolutionary events]

Hoffman visited South China and eventually found the geese in the rural villages of Swatow. In this part of China it is relatively warm in winter and there is surplus food to grow the geese and over-winter the stock. The birds are big, weighing up to 26lb (12kg), and are just like the USA Africans in size, colour and posture. Small flocks of birds grazed the roadside pastures, and he estimated that as many as a million goslings each year were sent north to Shanghai and Pekin for rearing. Hoffman's photograph confirmed it: the Tse Tau or 'Lion Head' is the 'outrageously misnamed African'. He also found a reference to the letter dated 1852 (*see* above), confirming that such large geese had been

*USA: the Hong Kong or China goose (left), and Col. Jacques white 'Bremen geese', imported shortly after the Sissons' Bremen geese (*see *Chapter 9) (Dixon and Kerr, 1853).*

exported to America from the province just north of Swatow.

X-W. Shi et al. (2006) also carried out an examination of the mitochondrial DNA of Chinese and European breeds of geese:

> All Chinese breeds and their maternal hybrids except the Yili breed showed an identical haplotype, named haplotype I or the Chinese haplotype; the European breeds and the Yili breed showed another haplotype, named haplotype II or the western haplotype. None of the haplotype found in the Chinese type was detectable in the western type and vice versa . . . Using 2% substitution per million years calibrated from the genera *Anser* and *Branta*, the two domestic geese haplotypes were estimated to have diverged approximately 360,000 years ago, well outside the 3000–6000 years in domestic history. Our findings . . . support the dual origin assumption of domestic geese in the world.

Although American literature frequently mentions the African in the early twentieth century, the breed seems to have virtually disappeared from Europe and had no standard here until 1982 when the American standard was adopted. The birds have grown in popularity as there have been imports in the last four decades, some of which have been mingled with the stock of earlier breeders such as those of Reginald Appleyard from the 1930s.

In addition to the Brown (Grey) African, which is the normal or wild-colour of the Swan goose, there are also the much less common White and Buff varieties. White Africans are very similar in appearance to the Russian white, knobbed Kholmogorsk. Buff Africans have been developed in America by Dave Holderread, and Blue Africans have also been bred in the UK, the blue gene arising from the Steinbacher. The neck feather partings inherited from the American Buff and Blue Steinbacher have introduced such faults in the African, which should be avoided.

## The Kholmogorsk

The Kholmogorskaya/Kholmogory/Cholmogorsk is an old Russian breed, but seems rather different in temperament and type from the Russian fighting geese (*see* Chapter 15), and is much larger. Phenotypically it is a white 'African' with a knob and dewlap which develops with age. The birds are said to be quiet and amenable, and will get on in large flocks. Numbers of 19,680 and 18,500 were cited in the USSR in 1974 and 1980 respectively.

The website www.komovdvor.spb.ru/en/gusi/xolm.html gives Abozin 1885 as their first literary mention. They are illustrated in white, grey and 'skewbald', similar to the illustration in Dürigen 1921. A weight range of 16.5–26lb (7.5–12kg) is cited. Twenty to twenty-five eggs are produced each year, but the best geese lay up to fifty, the eggs weighing 5.6–7.8oz (160–220g).

Dürigen (1921) found that although the goose had the same name as the ancient cattle-trading town of Kholmogory, near Arkhangel on the White Sea, it had nothing to do with this North Russian town and surrounding area; rather it was bred in the central or lower Volga region: 'This would correspond with its name of Saratowski [of Saratov], under which I found two examples of this high-beaked goose in the agricultural museum in St Petersburg.'

Dmitriev and Ernst (1989) also state that this breed was formed in the Central Black-Earth zone of the Russian Federation by crossing local white geese with Chinese. It was also said to be widespread around Belgorod, Kursk, Moscow Vladimir and Arkhangelsk (near Arzamas), amongst other regions. The live weight of the Russian adult birds is 15–17.6lb (7–8kg) in males, and 12–13lb (5.5–6kg) in females, but 19.8 and 15lb (9 and 7kg) respectively could be achieved – that is, close to the weight of the African. The Russian report suggests that the Kholmogory was removed

## DESCRIPTION OF THE SHITOU/AFRICAN

*Details of the Shitou. Distribution: Guangdong Province. Population size: 600,000 geese in 2002. Adult weights: 19.5lb (8.85kg) for the gander, 17.3lb (7.86kg) for the goose. Eggs: twenty-four per annum in year one, weight 6.2oz (176g); twenty-eight per annum in year two, weight 7.7oz (217g). Photographs: Dr Hui-fang Li. Details of the breed in China were published in* Fancy Fowl *(M. Ashton, 'African Geese', 2008).*

**Head:** Large with a stout bill. A broad, forwardly inclined knob protrudes from the front of the skull. Eyes large and bright. A large dewlap, crescentic in shape, hangs from the lower jaw and upper neck.

**Neck:** Long, massive, nearly the same thickness along its length. Slight arch.

**Body:** Large, long, and nearly the same thickness from back to front. Breast full and round. Breast and underline free from keel. Paunch clean or dual-lobed. Stern round and full. Tail held high, especially in ganders.

**Legs:** Thighs short and stout, shanks medium length, stout.

**Plumage:** Tight and sleek except on the neck where it is like velvet.

**Carriage:** Upright, especially when active; 30–40 degrees above the horizontal.

**Colour of the Brown (or grey):** Dark brown stripe running from the crown of the head down the length of the neck contrasts with the cream throat and dewlap. Outline of off-white feathers around the bill in mature specimens. Darker brown smudge along the jaw line. Pinkish-buff blush on the breast fades to pale buff, then white on the paunch and stern. Sides of the body and thigh coverts ashy brown, each feather edged with lighter shade. Back, wing bow and coverts similarly marked; primaries and secondaries brownish grey.

**Bill and knob:** Black.

**Eyes:** Dark brown.

**Legs and feet:** Dark or brownish orange.

**Weight:** 1997 standard weights, 22–28lb (10–13kg) in males, and 18–24lb (8–11kg) in females, match the weight recorded by Hoffman in China and current exhibition specimens here. Previous standard weights were 20lb (9kg) and 18lb (8kg) for male and female.

**Breed defects:** Lack of dewlap; lack of knob; white patches amongst coloured plumage.

**UK:** Standardized 1982 in the Brown (Grey) and White; 1999 in the Buff.

**USA:** Standardized 1874 in the Brown (Gray); 1987 in the White. *Status:* Watch (ALBC).

*Kholmogorsk in quarantine in Holland, imported from Russia, 2003. These are, phenotypically, White Africans. They lack the neck-feather partings of the Greylag - Hans Ringnalda. Peter Jacobs*

*Kholmogorsk (Dürigen 1921). Although Asiatic in type (African/Shitou), these birds show evidence of cross-breeding with Western geese, as in the European 'spot' pattern. The colour pattern is similar to 'Chinese' geese from Madagascar, illustrated in* National Geographic *Vol. 32, No. 4, 1967. Asiatic birds probably reached Madagascar with the settlers who sailed from South East Asia hundreds of years ago.*

*Head of an African gander.*

*A pair of exhibition Africans.*

*White, African-style dewlap geese, introduced to the UK in about 2008 from Eastern Europe. Their voice is similar to the Chinese, and they are lighter in weight than the African/Kholmogorsk. Similar Russian Lindovskaya geese were advertised recently in Germany. Egg production is 50 per cent higher. There are lightweight and heavyweight strains. Paul Jones*

from commercial farms and replaced by more productive birds, because it was characterized by low egg and gosling production.

One suspects that the breed has been made and remade, depending on the productivity of the birds. The White is scarcely obtainable in Britain, but smaller, African-style white geese with dewlaps have arrived in the UK circa 2008.

## Choosing Breeding Stock: Type and Colour

Quality Africans are impressive, massive birds with a stout bill, thick neck and large body, but active behaviour. They are tall, and outreach the top of 36in (90cm) show pens. The knob and dewlap take over a year to mature, and the birds only achieve full size at two to three years old.

Avoid squat or thin-necked birds; height is needed, but not at the risk of a fine neck which is indicative of the Chinese type. Carriage should be 30–40 degrees; the birds should be active, and the tail turned up. Choose birds with a tight breast and underline; a keel is not wanted. Even though the original standard demanded little development of the paunch, most birds have this feature. Make sure that the paunch is dual-lobed: this is more likely to be so in ganders than geese, and in specimens which are not too fat. The birds should have a large, broad head and knob (as wide as the head) and a semi-circular dewlap, but these features do develop with age.

Colour is also an important consideration. Cross-breeding with white Africans in the past has spoiled some Brown lines, so if there is obvious white as a blaze on the breast or white flights, these birds should be avoided. In Brown Africans, avoid birds with orange in the bill, orange around the eye and matching bright orange feet, as these colour points indicate a white ancestor – although orange in the knob can come with age or with frost damage.

## Eggs and Goslings

As with all geese, Africans vary in their laying capabilities; figures from China are from twenty-four to twenty-eight eggs. The hatchability of the average African and Chinese eggs in incubators is good as compared with Pilgrims and Brecons; the goslings should be quickly on their feet, and be strong, rapid growers.

The egg weight of 7.7oz (217g) quoted by Shi (2006) from China is interesting. Mature Africans do lay large eggs, and the emphasis on large birds for exhibition has made the egg size even larger. However, the largest 8.5oz (240g) eggs do not hatch well at all, having difficulty in losing sufficient water. 'Hypertyping' (*see* Chapter 13) is not in the interests of conserving a breed. Hatching time has been extended to thirty-four to thirty-five days, compared with thirty to thirty-two days in the 1990s, and the goslings often retain too much water. The largest goslings take two to four days to get on their feet. However, they are very

MANAGEMENT TIPS

* Unlike some other breeds of European geese, African geese never fly. They are too heavy and the wrong shape.
* The parent goose can be a good sitter but should not be allowed to hatch the goslings for fear of trampling them.
* Hand-reared, the African can be one of the tamest and mildest-mannered of geese, especially in the heavier exhibition strains.
* Africans do have a loud trumpeting voice; though not as frequently vocal as the Chinese, the volume is even greater.
* Both African and Chinese geese can become arthritic at around twelve to fifteen years of age, earlier than European geese. They are best housed in very cold weather.
* Life expectancy is less for the females, which become very heavy and may prolapse in the breeding season (with laying).
* Protect Africans from heavy rain as they can suffer from wet feather as they age.
* The Asiatics do have slightly different feeding habits in the wild. The Swan goose prefers wet areas for feeding, rooting out waterplants. Asiatic geese are also said to dive for insects in the water, which seems likely: Africans will chase daddy-long-legs and appear to eat them, a practice not followed by the other geese.
* Adult Africans suffer more than most geese from gapeworms, which impede their breathing. Like the Chinese they tend to pull at roots, and may get infected from the soil this way. If a bird is holding its neck at an awkward angle or making noises in its throat and windpipe, then worming is essential.

fast growers, and by three to six weeks have enormous legs which have grown at a disproportionate rate.

Production Africans lay 6.4–7oz (180–200g) eggs which are easier to hatch but do not produce the 'super African'. UK stock tends to produce fertile eggs only early in the season, and a surplus of males.

## Chinese Geese

Chinese originally came in two colours – White and Brown, also called Grey or Fawn. The breed was introduced early to the Western world, and 'China' is said to have been an early preferred name. The Brown variety is a direct descendant of the Swan goose, which has identical colouring and feather texture. The Asiatics lack the Greylag feather partings on the neck – instead the plumage looks like velvet. The beak in the Swan goose is black, as it is in the Brown Chinas, but it does not exhibit the raised knob; this feature seems to have developed after the domestication of the bird. The knob is covered in soft skin and is warm to the touch.

Robinson (1924) states that the earliest record of Chinese geese in America is in the correspondence of George Washington, and was first published in Haworth's 'George Washington – Farmer' in 1915. It discloses that in 1788 he received two Chinese pigs and with them 'a pair of white Chinese geese, which are really the foolishest geese I ever beheld; for they choose all times for setting but in the spring, and one of them is even now (November) actually engaged in that business.' Robinson thought they were white, as the colour was well known in Virginia early in the nineteenth century.

Chinese must have been in Britain prior to 1845 (*see* the d'Hondecoeter painting at the beginning of this chapter) because of the inclusion of the 'knobbed or Asiatic' geese in the first national poultry show. A detailed reference to them in Dixon (1848) stated that 'Mr Alfred Whittaker of Beckington, Somerset, owned a flock of white Chinese which were from imported parents, and were hatched on board ship from China.'

Nolan (1850) also recorded the white China: 'It is a beautiful variety, next in size to the above [the African] and approaches nearest to the swan of any other goose. It is snow white, knobbed on the beak with orange legs, and truly ornamental on a sheet of water.' He was more dismissive of the Black-legged Chinese goose, which was smaller than the white, but marked like the Hong Kong (African).

### Growth of Popularity in Britain

Chinese geese may not have been very popular in Britain from the 1850s to the 1920s, with complaints that they scarcely bred; they only merited a couple of sentences under 'Other Breeds' in the 1922 Poultry Club Standards. The Brown variety was recorded as having black legs (a characteristic which still exists in Australia), though the Americans had already plumped for 'dark or dusky orange' by 1905.

There seems to have been little interest in showing them, but there were some commercial flocks. These were featured in *The Feathered World* (1928, 1930), the articles highlighting the exceptional egg production – up to 110 per annum – from American utility birds. Appleyard (1933) also kept the birds as a good utility line. His geese often laid fifty to sixty eggs a year, and instances of 100 were not unknown

By 1954 a complete standard had been advertised

in *Poultry World* and published in the standards book. The birds were popularized in the 1950s by Colonel Johnson. His birds, which won the top awards at the time, were a large, dual-purpose type, long in the body and without a very pronounced knob. They had the reputation of being prolific layers, the White variety often seeming to have the edge for egg production over the Brown.

The preferred Chinese type for exhibition today is a smaller American strain. The ganders weigh 10–12lb (4.5–5.5kg), the females 8–10lb (3.5–4.5kg). The Americans selected a light-boned, short-bodied bird with a view to producing a small but economic goose for the table. This refined goose, with elegant carriage, came to Britain in the late 1970s. They are beautiful to watch in action, being graceful on land and water, and when hand-reared they have a good temperament.

*A group of showy young Chinese with plenty of reach, and stylish carriage.*

DESCRIPTION

**Head:** Medium in size. Bill stout at base. Knob large, round, prominent (larger in the gander than the goose). Eyes alert.

**Neck:** Long and slender, carried upright, with graceful arch.

**Body:** Compact and plump. Back reasonably short, flat, broad and sloping to give characteristic upright carriage. Well rounded, plump breast. No keel or dual lobes; single central fold in laying geese. Wings large, strong and carried high. Stern well rounded. Tail carried high in American exhibition-strain ganders where the wing tips cross over in front of the tail.

**Legs:** Medium length.

**Plumage:** Sleek, tight and firm excepting neck, where it is soft and without furrows.

**Carriage:** Upright, body held at 45 degrees when active. Ganders especially should stand with their heads and tails held high.

**Colour:** As in the Brown African and White African.

**Breed defects:** Lack of knob. Presence of dewlap. White patches in coloured plumage.

**UK:** Mentioned in 1922; full standard description 1954.

**USA:** Standardized 1874: Brown and White.

**Status:** Watch (ALBC).

**New colours:** Blue and Silver-blue, but they have not been standardized.

*White Chinese geese – C. Marler.*

## Choosing Breeding Stock: Type and Colour

The right type of stock will depend on its purpose. The heavier English types are better layers, the lighter American types are the show birds and have a better temperament. In America, exhibition birds average one or two pounds less than commercials and are not penalized for this lack of size in the show pen, with shape being more important.

It takes up to two years to develop the perfect head shape, with a well rounded but forward-inclined knob, as shown in the specimen in the 1982 UK standards book. It is therefore important to see the stock birds if buying youngsters. Ideally, birds of two to four years old would be bought, but generally it is only youngsters that are available.

Show birds will gain many points for a lively, elegant carriage. The best exhibition birds have an upright stance and hold their heads high when excited. This posture will exaggerate their reach. A gracefully curved neck should combine with a short, deep body.

In colour requirements, follow the advice as for the Africans. White Chinas are crossed with Browns, but this spoils the colour characteristics of Browns, which acquire white under the chin and even a blaze of white across the breast, and bright orange feet. They can come out self-coloured, nevertheless, the Chinas not carrying the 'pied' gene, unlike European breeds. Birds heterozygous for white may look brown, but will breed 25 per cent whites.

'White' in both Chinese and African geese is caused by a single, recessive, autosomal 'c' gene. When homozygous for c, the birds look white because this colour masks the wild-colour of the Swan goose. Birds which are heterozygous for white sometimes show little evidence of the gene, but from a pair of such birds, a quarter of their offspring will be homozygous for white. In European breeds 'white' is caused by the interaction of the recessive gene for 'spot' (*sp* [pied]) and the dominant sex-linked 'dilution' gene (*Sd*). These genes are absent from pure Chinese and African stock (*see* Chapter 7).

## Eggs and Goslings

Chinese can lay from January until June, without many breaks. Figures of seventy-five to a hundred eggs have been achieved, though this is only from young geese. Old stagers settle down to a much more reasonable thirty eggs, and American show strains are unlikely to be prolific, even when young. However many eggs Chinese (and Africans) lay, they do not seem to lay double-yolkers.

The eggs are easy to hatch in the incubator as long as the parents are not too closely related. The eggs readily lose water, allowing the air sac to develop well, and the gosling to hatch. These eggs do need higher humidity than eggs from most other breeds.

Chinese need not be expected to go broody, but if they do, the goose is less likely to crush the eggs than the larger breeds. The ganders are excellent fathers and love to guard the goslings, and even adopt ducklings in the breeding and rearing season.

### MANAGEMENT TIPS

* Unlike other light breeds of European geese, Chinese geese generally do not fly. The body is plumper and more compact in proportion to the wings.
* These are small geese and so require less supplementary feed than other breeds.
* The big advantage of having Chinas is in the garden: they will keep the grass down and are not nearly as destructive of trees as Brecons and Embdens. Unlike the Greylag derivatives, they tend not to chew twigs and saplings.
* Both Chinese and Africans also like to pull up creeping stems of grass and buttercups and dunk a mouthful of grass and mud in a bucket. This can make the lawn patchy.
* Chinese were used as weeder geese for crops in the USA and are excellent for this purpose in a garden or tree nursery.
* Chinese are easy to sex: the geese can usually be sexed by voice by sixteen weeks. The ganders are also invariably taller, have a fuller developed knob (though this takes two years to reach its full size) and have a strident voice in a single note, quite different from the females which give a very distinctive 'oink'.
* Like the Africans, they have a shorter life expectancy than European geese, and are more prone to arthritis.
* Chinas are marvellous watchdogs. They are sensitive to the slightest noise, and once spoken to, cannot resist shouting back. Dixon (1848) commented: 'Any fowl-stealer would be stunned with their din before he captured them alive, and the family must be deaf indeed that could sleep through the alarm thus given.'
* If you have immediate neighbours this watchfulness is a nuisance, especially if the birds are out in a fox-proof pen at night. In a flock, ganders tend to get involved in minor scraps, and the females like to egg them on. Therefore shut them up at night, to keep them quiet.

The goslings are great fun to hand rear. They are light and strong, and quick on their feet. They are great talkers if kept in the kitchen for the first week, and benefit from being handled. They become extremely tame and are likely to remain so as adults, especially if they are an American strain, when it is rare to get peevish birds.

## Varieties of Geese in China Today

The 'African' (Lion Head, or Shitou) and 'Chinese' are two breeds of geese from China recognized in the West. Published in 2003, the FAO paper 'Report on Domestic Animal Genetic Resources in China' reveals no fewer than twenty-four types of geese native to China.

Biological diversity is essential for maintaining agricultural production and food supply, yet the increasing scale of producing units, and the technology used in food production, both tend to limit diversity. Recognizing this, United Nations FAO (Food and Agricultural Organisation) involved more than 150 countries in signing the *Pact of Biological Diversity*.

The long-term evolution of farm stock has meant that local varieties are often best suited to efficient food conversion in their local setting. So it is important that such traits are preserved in the global gene pool in order to have the resources for increasing agricultural productivity, in a world of dwindling resources and increasing human population. Countries were required to list their animal genetic resources (AnGR). Further information can be found in the FAO 2003 report and in papers published by Zhu et al. (2010) and X-W. Shi et al. (2006).

# 9 White Geese

## Introduction

White geese are descendants of the Greylag and include the Embden, Roman, Bohemian and other European white geese. White geese are the commercial favourites: from goose down in Roman times to the goose-down duvet today, white is the luxury product. For centuries, geese have provided not only meat but also feather and down. These products were, and still are, of real economic importance, and as a result there are numerous white varieties which are very similar to each other; out of their regional context, these varieties often cannot be specifically identified.

Edward Brown observed white geese across Europe in 1906 in his book *Races of Domestic Poultry* where he described the Roman, Embden and Danubian (Sebastopol). In his 1912 report on the German poultry industry, he noted great differences in size between the geese that were met with in northern and southern Germany. Throughout south Germany, Austria and Hungary, the geese weighed only 8–10lb (3.6–4.5kg); they were of Roman or Italian type and descendants of the birds that saved the Roman Capitol.

In the richer grazing lands to the north he said that the product was 'much finer flesh' and weighed 15-20lb (6.8–9kg). Leading goose breeds were, of course, the **Embden** and the **Pomeranian** (which is a type of goose in grey, saddleback and white). In addition there was the **Diepholz**, described by him as an off-shoot of the Embden, bred as a winter layer and coming into profit in September, so goslings could be marketed in early spring.

*Diepholtz goose with a small, neat head, at Sinsheim, 2003. These are much smaller than the Embdens, weighing 12–13lb (5.5–6kg). They are a bit larger than Roman geese in the UK, illustrating the gradation in type and size in regional white varieties in Europe.*

There are several European strains of commercial white geese. They are all smaller than the massive Embden, fast growing, and have a better flesh-to-bone ratio for the Christmas producer. For example, **White Hungarian geese** (cited by *FAO, United Nations Food and Agricultural Organisation)* weigh 10.4–12lb (4.7–5.5kg).

**White Italian geese** are also used commercially in Europe. They were mentioned as an import to the UK in the late nineteenth century in the magazine *Fowls for Pleasure and Profit* (1888). Although the birds did not compare favourably with Embdens for size, as layers they were said to have no equal. These fast-maturing birds made 9–10lb (4–4.5kg) weight at three months, and of the white and grey varieties available, the white was said to be the superior. These birds were too large to be the typical Roman geese, as they grew on to up to 16lb (7.3kg). Wright (1902) also wrote of:

> a variety . . . highly recommended about ten years ago under the name of Italian geese. It has been

> stated to be unusually prolific, laying 50-60 eggs in one laying and sometimes a second. Mr Tegetmeier describes them as mainly white, with a blue-grey head, a grey roundish spot between the shoulders, and grey thighs. But a great many we have heard of have not come up to that standard, and have been decidedly small.

FAO reports that Italian geese have been under genetic selection since the 1960s at an experimental station in Poland. Males weigh up to 15.4lb (7kg) and females up to 14.3lb (6.5kg). Annual egg production for a line designed for high egg laying is sixty to seventy eggs; egg weight is 5.6-6.3oz (160-180g).

The **Danish Legarth** goose is a strain which has also been tailor-made to have a rapid growth rate with maximum feed conversion benefits. A Danish supplier cites a growth rate of 13.9lb (6.3kg) at sixty days and 15.9lb (7.2kg) in ninety-eight days, with feed conversion rates of 1:2.6 at sixty days and 1:4.5 at ninety-eight days. Intensive poultry production demands that birds are slaughtered at a young age to get the greatest cost benefit from inputs (*see* Chapter 21).

In France, geese are still of great economic importance, and there are several types of white geese which are considered to be standardized breeds in France. They include the **Poitou** (13-14.3lb/6-6.5kg), **Touraine** (11-15.4lb/5-7kg) and **Bourbonnais** (15.4-22lb/7-10kg).

*Emporda goose, a white, crested Spanish breed, also standardized in Germany.*

Standardized exhibition white geese include the Embden, Emporda, Roman, Bohemian and Czech. Although the Embden, Czech (or Bohemian), Diepholz and the Deutsche Legegans (the egg-laying goose) have been standardized in Germany, other countries tend to be more conservative in their choice of white exhibition birds, which need to be distinctive. Both the UK and USA initially standardized only the Embden and Roman; the UK has recently (2008) also admitted the Czech (now known as the Bohemian).

## The Embden

### History of the Embden in Germany

There is a general - incorrect - belief that any white goose is an Embden. The Embden is one of Europe's more ancient breeds, and the standard is jealously guarded in Germany. The birds have been exported to the rest of Europe, Australia and the USA, but the true type remains in its country of origin, for there have been adaptations of the breed elsewhere.

The Latin name of the Embden is *Anser domesticus frislandicus*. The breed comes from the area of Ostfriesland in northern Germany, and its initial development may be as much as 2,000 years old. Even in the first century Pliny Secundus (c. AD40) pointed out that white plume and down were preferred to grey (*see* Chapter 1).

German references (Bruno Dürigen, Jürgen Parr and Horst Schmidt) indicate that systematic breeding certainly took place in the thirteenth century. Feathers and goose meat were exported from the valley of the River Ems, where large flocks were kept. However, the floods of 1863 affected their numbers badly, and economic use of the breed almost came to a halt. Fortunately the birds survived in stock exported to Hungary and Czechoslovakia, where some flocks had been kept pure. Others were crossed with local breeds, because the Embden was of course being used to improve other regional types of commercial geese.

Edward Brown thought that the early twentieth-century Embden was the result of crossing the German white and English white geese. He noted that many English Embdens were exported back to Germany, so the stock on both sides of the North Sea became closely allied. Parr considers that this was a retrograde step, the English type of bird spoiling the elegance of the original. German breeders, he says, complained about the large head, thick neck and Toulouse characteristics such as dewlap and breast keel.

*Embden geese. Birds are judged on points at these German shows, which can run for four days, and give judges a lot more time to assess the birds.*

*A typical German Emden gander in 1882, as compared with the English style showing Toulouse characteristics (Dürigen, 1921).*

Of course World Wars I and II meant that little attention was paid to the breeding of the birds in Germany, and it took until 1960 to regain the typical look lost during the war. A big show line-up in the 1990s included up to fifty of these birds, as compared with only a few individuals at a British show.

## The Embden in the USA and UK

These birds were introduced into the USA in 1820, as indicated in a letter from Mr Sisson in *New England Farmer* in 1825 (*see* Robinson, 1924):

> In the fall of 1820 I imported from Bremen (north of Germany) three full-blooded, perfectly white geese . . . Their properties are peculiar; they lay in February and set and hatch with more certainty than the common barnyard geese, will weigh nearly, and in some cases quite twice the weight, have double the quantity of feathers, never fly, and are all of a beautiful snowy whiteness . . . I have one flock of half-blooded that weigh on average when fat 13 to 15 pounds. The full-blooded weigh 20 pounds.

The term 'Bremen' was also used by Col. Samuel Jacques, who received his geese in Boston, 1821. Note that both American authors stressed that the Embden goose was white, in contrast to the farmyard grey goose they were used to. The Bremen name continued to be used in America until about 1850, when some writers started to use English descriptions. The main city of the source region is of course Emden, the English inserting a letter 'b' into the word – but when the USA standard was published in 1874, the term 'Embden' was used as the official name. The name was recorded in Britain when Moubray (1816) said that 'At present, 1815, the Embden geese are in the highest esteem. They are all white, male and female, and of a very superior, indeed very uncommon size.'

Embdens were referred to briefly in E. S. Dixon (1848) and in Wingfield and Johnson (1863): 'Beyond their great size, the uniform clear white of their plumage, we are at a loss for any sign of specific differences between these and the ordinary goose of our commons.' Quoted in Tegetmeier (1867), Mr Hewitt said:

> In Geese I must claim the pre-eminence of the Embden variety. I have traced the best specimens of this kind through several owners, and found that the originals (in these instances) came from Holland. One of their great advantages is this, that all the feathers being perfectly white, their value, where many are kept, is far greater in the market than is ever the case with coloured or mixed feathers. In weight, too, these birds have an advantage even over the Toulouse.

In Britain this desire for the heaviest birds at the Victorian shows resulted eventually in weights of 32lb (14.4kg) for ganders and almost 26lb (11.8kg) for geese being quoted in Lewis Wright, and over 30lb (13.6kg) for males in Brown 1906. This size was achieved by crossing Embdens with Toulouse, as was indicated by a quote by the Amhurts in Wright, 1902. This cross affected the 'type' of the birds, and also resulted in an increase in egg production from fifteen to thirty eggs per year. The English Embden acquired the deep beak and strong head, set on a long but fairly thick neck. In contrast, the German birds were finer but showed less propensity for a keel and dewlap, which can still be a recurrent problem in the bigger English Embdens bred today.

'English-style' Embden stock has become extremely difficult to obtain in the UK because few people really understand what an Embden is. It is not just a white goose: it has to have the correct symmetrical shape and stance, and be massive. Such birds, which reach 34lb (15.4kg), are not produced by commercial goose producers because they come from large parents where the gander may not be particularly fertile, and the goose lays fewer than thirty eggs.

Exhibition Embden geese have only ever been produced by very few breeders in the UK. In the 1970s, the famous Oliver line produced show winners, and these continued to be exhibited into the 1980s by Jacob Lory. Vernon Jackson's line was also illustrated in the *Waterfowl Standards* of 1982 and became the type to emulate.

There are probably two reasons for the paucity of excellent stock. Firstly the demand for the breed has never been particularly high, except for use as a commercial cross, so breeders have not been encouraged to produce many of the birds. Relatively few are sold, and even top quality birds do not command a particularly high price for the amount of work involved. Secondly, the English exhibition birds became increasingly inbred in the 1990s. We found some of the goslings very difficult to rear. The eggs were fertile, they hatched well, but the largest goslings were distinctly unhealthy and failed to thrive. The smaller ones, which had obviously inherited different genetic material, were fine and grew, but not into exhibition birds.

To revitalize the breed, Tom Bartlett imported tall German birds in the 1990s, which reached nearly one metre in height. They have a long body as well as a long neck, a trim dual-lobed paunch and no suggestion of a keel. Perhaps significantly, Parr demands that the German Embdens show no trace of 'a hump between the bill and the forehead', which would suggest Chinese blood. The reduced feather partings on the birds' necks do suggest that Asiatic geese may have been used at some point in the past to regain the German Embden look. The head is also different from the English Embden: it is rather slim, and when viewed from the side gives the impression, together with the beak, of being oval. The bill itself has a straight culmen.

*Embden gander (the property of Miss M. E. Campain): winner of First Dairy 1906 and 1908, Second Dairy 1907, and several first and other prizes at Birmingham, Manchester etc. (*The Encyclopaedia of Poultry, *Vol. 1, J. T. Brown, 1909). This is still a desirable type of English Embden, similar to the type exhibited by Captain Harrop at the Dairy Show in 1932.*

DESCRIPTION OF THE TRADITIONAL ENGLISH EMBDEN

**Carriage:** Upright and confident.

**Head:** Strong, bold; stout bill.

**Neck:** Neck long, well proportioned, without a gullet.

**Body:** Broad, thick and well rounded. Back long and straight. Breast round. Shoulders and stern broad. Paunch deep, dual-lobed. Large, strong wings.

**Legs and feet:** Legs medium length. Shanks large and strong.

**Plumage:** Hard and tight.

**Colour:** Pure white.

**Bill:** Orange.

**Eyes:** Blue.

**Legs and webs:** Orange.

**Weight:** Males 28-34lb (13-15.5kg), females 24-28lb (11-13kg). (**USA:** Males 26lb/ 11.8kg, females 20lb/9kg; **Germany:** Males 24-26lb/11-12kg, females 22-24lb/10-11kg.)

**Breed defects:** Uneven undercarriage, indication of a keel or dewlap.

**UK:** Standardized 1865. Status: Rare breed.

**USA:** Standardized 1874. Status: Not endangered - primary production (ALBC).

## Choosing Breeding Stock

The first obvious requirement is a good size, but there are also other points to look for and avoid. Overweight birds may develop a slight dewlap under the bill or keel on the breast; Embdens should not have these features, so also reject any lightweight birds that have them. Ganders should be white, but young females may show a certain amount of grey on the rump as a sex-linked feature; however, this usually moults out when the birds are in their second feathers by sixteen to eighteen weeks. The preferred colour of the bill and feet is orange in Britain, but some birds show pinkish extremities, a characteristic of East European geese. Although this is not desirable, it is a minor point.

Although Embdens are expected to be large, the birds must be able to walk comfortably. They should stand upright and alert when approached, and not dip at the breast. Always move the birds around carefully to make sure they do not limp. If they are overweight they should be slimmed down a bit before the breeding season, particularly the gander. Ganders must have a symmetrical dual-lobed paunch, viewed from the back and front. This is preferable in females too, but less easily achieved.

### Breeding Embdens

It is essential, as with Roman geese, to understand something of the breed by visiting breeders and shows before buying, otherwise you are likely to end up with an average white goose. You may be lucky to come across a flock of Embdens which have originated from a pure line, but expertise is needed to grow these birds up to exhibition size, and size is soon lost over a couple of generations without good management.

Embdens begin to lay in mid February, laying up to thirty eggs in the season, usually in two clutches. The largest geese have a tendency to lay some double-yolked eggs, but fertility is not a problem using young ganders weighing about 26lb (12kg). The females, unlike the Toulouse, are quite keen to go broody but are too heavy to hatch goslings successfully, even in a very large nest with easy access. The eggs are more hatchable than most in incubators, the air sac achieving the correct size for hatching, where the Brecon and Pilgrim eggs fail due to insufficient water loss. Sisson's comment about them hatching with more certainty than the 'common barnyard goose' remains true.

### Rearing Goslings

On hatching, Embden goslings need a lot of care. They are large, much bigger than Toulouse, and it is essential to place them on a rough surface where they can get a good grip with their feet, otherwise their legs will spraddle (*see* Chapter 20).

Embdens, as well as other European breeds of white geese, are auto-sexing. Males and females typically have saddleback markings and a coloured head in grey fluff, the males being paler in these areas compared with quite dark grey in the females.

The goslings are the fastest growers I have ever seen. For this reason they cannot be kept in a mixed bunch after about a week because the smaller goslings may get crushed. By three to four weeks they have enormous feet and thick legs, which are necessary to support their weight. However, it is important not to let them get too obese: they readily get rolls of fat on the breast at this stage, so it is probably better to cut the food back a bit at this point, particularly the protein. Embdens need exercise and grass rather than sitting down to a bowlful of pellets all day. Apparently in Yorkshire there was a tradition of 'walking the geese', a task for the children when they came home from school, and this undoubtedly served to strengthen the birds. A supplement of brewers' yeast in the diet also helps.

## Roman Geese

### History

Everybody loves small white geese. Petite geese are garden friendly, lay lots of eggs, and have been widespread in central and southern Europe for over 2,000 years.

The Roman goose is famous for its part in saving the Capitol at Rome during the siege by the Gauls in the fourth century BC. Livy and Lucretius and other ancient writers mention both white and coloured geese. Brown (1929) still distinguished two varieties: the Roman and Padovarna, he said, were evidently of one race, not distinct breeds, differing only in feather colour. The Roman was pure white, but in the Padovarna the head, the back of the neck, the wings and sometimes the back were grey.

Brown concluded that this white domestic goose could have been disseminated throughout Europe by the Roman army and via commercial dealings between the conquered lands and Rome. He felt that the small white goose found on his travels throughout South Germany, Austria, Hungary and the Balkan States was probably descended from the Roman, the oldest breed of all.

In contrast, Robinson (1924) pointed out that Italy was a receiver of geese from central Europe, and that the process of domestication of the Greylag was much more practicable in the regions where it bred – around the mouth of the Rhine, in Pomerania and in Scandinavia. It is likely, of course, that domestication took place at several points quite independently across Europe. For example, he notes that when the Romans invaded Britain in 55BC, they found tame geese there because Caesar reported that they were kept only for sport.

Whatever the origin of these small white birds, they seem to appear first in Italy in the written record.

### The Roman in the UK and the USA

According to the *British Poultry Standards (1954)*, the Roman was introduced into Britain in about 1903. Brown (1906) described them as follows:

> The body is long and broad with a well developed sternum; breast broad yet not very prominent, thus resembling the Embden; back broad; neck long, fine and arched; head fine, in some specimens with a small crest at the back of the head; eye blue, surrounded with bare skin of pale yellow; beak very thin and short, nearly rose in colour, with a tip of ivory white; wings large, not carried far back; legs strong, reddish grey in colour, with white toe nails; the general appearance is a well

developed goose, not so upright or large as the Toulouse or the Embden. Weight: males 10–14 pounds; females 8–12 pounds.

In the USA the weights of the birds are quoted at 12lb and 10lb (5.5kg and 4.5kg) for the gander and goose respectively. Emphasis is put upon the fine bone and plumpness of the birds, this shape being complemented by the straight, short neck.

There is also a tufted (crested) variety. This is unusual in geese, but a crest can appear in Pilgrims, Buffs and Sebastopols and otherwise unremarkable white geese. The crest is not as pronounced as in Crested ducks, and is merely a raising of the feathers, rather like a close-fitting helmet instead of a bobble. It is nevertheless produced by a deformity of the skull and is not of any advantage to the bird. Crested Romans can be found in Britain but are more popular in the USA.

DESCRIPTION

**Carriage:** Almost horizontal.

**Head:** Neat, well rounded. Bill short.

**Neck:** Upright, medium length, not thin.

**Body:** Compact and plump. Back wide and flat. Breast full and rounded, not carried high. Paunch dual-lobed, not too heavy. Wings long, strong. Tail carried horizontally.

**Legs:** Short, light-boned, set well apart.

**Plumage:** Sleek, tight and glossy.

**Colour:** Glossy white plumage.

**Bill:** Orange-pink.

**Eyes:** Light blue.

**Legs and webs:** Orange-pink.

**Weight:** 12–14lb (5.5–6kg) male, 10–12lb (4.5–5.5kg) female.

**Breed defects:** Oversize and excessive weight.

**UK:** Standardized 1954.

**USA:** Standardized 1977. Status: Critical (ALBC).

### Choosing Breeding Stock

Embdens and Romans are white, but not all white geese are pure breeds; furthermore there are many commercial white strains. Cross-mating using Romans, Embdens, Chinese, etc. is practised in order to maximize the benefits of the prolific egg layers, but also to add weight. The resultant cross-breeds are also easy hatchers and easy-to-rear youngsters. Many people have bought commercial white geese in the belief that they are Embdens, just as others have bought 'Romans' that were double their ideal size. If in doubt, buy white breeds only from a reputable breeder.

Temperament is important when choosing this breed. The 1982 standard described it as 'active, alert, docile, rather than defiant'. This comment on their demeanour applies to some strains but not others; indeed, we have had some very fractious Romans that were fierce to handle. However, breeders who have kept these little geese for years have generally found them tame and charming.

### Eggs

Romans can be marvellous egg-layers; Brown (1906) cites 60–110 eggs between October and June (which may have been achieved with extra lighting). E. A Taylor (*The Feathered World*, March 1916) also praised them as hardy birds, and great buttercup exterminators, consuming 2lb (1kg) of feed each weekly, living in the open, and averaging 100 eggs per annum.

A more realistic figure of forty-five to sixty-five eggs in a season was given by Appleyard, who found them to be the ideal small goose. His Romans slept in the open too, and even when given a shelter, preferred to live out in the snow, quite happy and contented (as long as they were protected from foxes). Because they are small, active geese, one gander can be kept with up to five females, and fertile eggs are still feasible. This is unusual for geese, which frequently pair or make a trio, making them a rather uneconomic proposition because a high proportion of males to females is required for breeding.

## Czech and Bohemian Geese

Roman geese have never been that popular in UK show pens because they are just small white geese. They are not as interesting in colour as Buffs or Pied geese – and what indeed makes a Roman? Is every small white goose a Roman, such as the feral birds which can be found 'dumped' on various lakes?

Central European breeders seem to have grasped this nettle, and have developed a special small white goose. Initially named the Czech (Tschechische) goose, it is immediately apparent that it is different. Weighing only 8.7–11lb (4–5kg), these little geese are the 'Call ducks' of the goose world. Egg production can average forty-five eggs with an egg weight of 5oz (140g). Small, chatty and with chubby cheeks, these birds have instant appeal and were included in the 2008 *UK Waterfowl Standards.*

In the 1950s, breeders developed a small white goose which could fend for itself on mainly grass, and provide a good meal for the table. The breast is rounded, and the abdomen carried clear of the ground on short, strong legs. A feature of the goose is its short, fairly thick neck with pronounced feather partings. The geese are busy little birds which do talk to themselves. Goose breeders have found them good sitters - and woe betide anyone who interferes with their brood! They are very defensive parents if they are near other birds. As garden birds we have found them to be sweet-natured and tame.

Rather confusingly, this 'Czech' goose developed in Germany has now been re-named as the Bohemian (2009). The birds typical of the Czech Republic have retained the Czech name. These are also small white geese, but a little longer in the body and shorter in the leg.

*Bohemian geese at Hannover Show, 2009.*

*Small, white Bohemian geese are single-lobed, broad-breasted birds. They are very similar in size and shape to the Oberlausitzer or Wendish goose (from Saxony, adjacent to Bohemia), illustrated in Dürigen 1921 in white, pied and grey. These small but powerfully built land geese are typical of this part of Europe.*

# 10 Sebastopol Geese

## Introduction

The Sebastopol has been called the untidy goose, or *unkempt* goose. We are often asked if the birds are in moult, and local farmers might enquire if the birds have been fighting, because by the breeding season the curled feathers may have become a bit sparse. The wind and rain of winter will have taken their toll, and a few feathers will have been tweaked out by bickering birds. But Sebastopols are spectacularly beautiful in the autumn when the volume of their new white feathers can look superb. Kept in good conditions, they can maintain their feather quality all year, and are a useful utility goose, too.

## Breed History

During his visits to Hungary and the Balkans, Edward Brown (1906) saw numerous geese with loose feathers in the lower regions of the Danube and in countries surrounding the Black Sea. Such birds had been brought to Britain from Sebastopol in the Crimea by Lord Dufferin, and they caused quite a stir at the Crystal Palace:

> These birds are somewhat smaller than those of this country at a mature size, but they are of the purest white . . . whilst the more conspicuous portion of their plumage is of a curly nature affording a very striking contrast to the feathers of the ordinary English goose. The feathers of the back are curled and frizzled upwards. The secondary feathers of the wings are elongated and twisted, also the tail coverts.
>
> *The Illustrated London News* 8 September 1860

Danubian was the preferred name at that time, but the breed was referred to as Danubio in Spanish and Italian, l'oie frisée in France and Lockengans in Germany. Brown thought that the breed had been developed from the ordinary white goose (the Roman type), where curled feathers had arisen as a mutation, since feathers and down were an important part of production in Eastern Europe at that time, and earlier.

The Victorian specimens very much conformed to an East European origin. They were small, 8-12lb (3.6-5.4kg), in other words with weights very similar to the Roman. Moreover their legs and feet were then rosy red, an inheritance from the eastern Greylag race.

Since these early imports, the West European and American birds have developed along slightly different lines. The standard weights have increased to 10- 16lb (4.5-7kg), and the bill and leg colour is often now more orange, although the UK standard asks for orange-pink.

DESCRIPTION OF THE WHITE CURLED-FEATHER VARIETY

**Carriage:** Horizontal.

**Head:** Neat, bill of medium length.

**Neck:** Medium length and carried upright.

**Body:** Appears round because of the full, loose feather.

**Legs:** Fairly short and stout.

**Plumage:** Only feathers of the head and upper neck smooth: feathers on lower neck, breast and remainder of body profusely curled. Back feathers should be broad, not wispy. Wing feathers should be free from stiff shafts.

**Colour:** Pure white preferred; traces of grey in young females allowed.

**Eyes:** Blue.

**Bill, legs and webs:** Orange-pink.

**Weight:** Male 12-16lb (5.4-7.3kg), female 10-14lb (4.5-6.3kg).

**Breed defects:** Wispy plumage on the back of the curly; rough/slipped wings.

**UK:** Standardized 1982 White, 1997 Buff. Status: Rare breed.

**USA:** Standardized 1938 White. Status: Threatened (ALBC).

*In 1860, the imported Sebastopols were smooth-breasted. (Illustration of the pair exhibited by D. Bayly, sent to him courtesy of John Harvey who had been cruising in the Black Sea.)*

## Type of Plumage

Sebastopols exist in two feather types, referred to as the curled-feather or curly-breasted and the smooth-breasted.

A curly pair (homozygous for the Sebastopol mutation *Sb* (*see* Chapter 7) will breed only their own form. Such birds have curled body feathers, and long trailing flight and back feathers. On the larger feathers there is degradation of the rachis (quill) of the flight feathers, and fluting of the vane. The elongated quill is split so that the feathers fall and touch the ground. This type of Sebastopol was standardized in the USA in 1938, but it took until 1982 to admit it to the UK standards, despite the breed's apparent introduction by 1860.

The *smooth-breasted* type is heterozygous for the *Sb* gene, which is incompletely dominant. A pair of birds heterozygous for *Sb* will breed 50 per cent of their offspring like themselves, plus 25 per cent curled-feather birds (homozygous for *Sb*) and 25 per cent plain-feathered - that is, lacking the *Sb* gene. The smooth-breasted types have long, curled, trailing feathers on the thighs, also falling from the scapulars and sometimes the wing coverts. In a good specimen, the scapular feathers lift from the back and cascade to the ground. The better birds usually have a fork in the tail feathers, indicating their potential to breed good feather quality. In 1997, the smooth-breasted was also admitted to the UK standard because of its popularity, and also because this is the only type recognized on the continent.

## Choice and Management of Breeding Stock

The gene affecting the feather structure behaves differently from the poultry frizzle gene where *one* such gene causes a 'frizzle' appearance with a reverse upward curve to the contour feathers. In its homozygous form, this gene causes the chicken to become 'woolly' (*see* Carefoot, 1985). Thus the standard frizzle chicken is heterozygous for the frizzle gene, and the standard curled-feather Sebastopol is homozygous for the *Sb* gene.

### Curled-feather Breeding Stock

Care should be taken in selecting the curled-feather parent birds. The best specimens should give the effect of a fluffy round ball of feathers, their long flight feathers trailing softly to the ground. It is best to select for birds with broad body feathers which give better weather protection on the back; all too often the back

*Sebastopol geese. Lori Waters*

*The curled-feather variety tends to have corrugated skin on the undersides of the feet (this is not bumblefoot). It probably goes with the Sb gene, so birds should be given good footing – grass and paddling water.*

plumage degrades into wisps, which should be avoided

Another problem with the curly variety is the way in which the wing feathers lie. In good specimens the shafts of the primary and secondary feathers, if examined closely, will be seen to be limp, because they are frequently broken along the length of the quill, meaning that the bird cannot fly. This broken effect allows the feathers to trail, but from this can arise two problems.

Firstly, the weight of the feathers can pull the final bone of the wing out of line, resulting in oar wing. Rapid growth on a high protein diet at three to five weeks of age can cause the weight of the blood in the quills to become far too heavy for the muscles to support, resulting in the distortion of the bone that will carry the primaries. Permanent deformity will ensue if this situation is not rectified immediately. This can happen in Sebastopols, although more commonly the wings can be normal at three to five weeks but fail to support the weight of the extremely long feathers later. It is prevalent in Sebastopols because the curled-feather type also lacks thigh coverts to hold the wing feathers in as they grow

Secondly, Sebastopols also seem to sprout new wing feathers during the autumn, and if the quills are in the short stage and protruding, then they can give the impression of oar wing, when there is nothing wrong with the bird, and the feathers will trail when given a few more weeks' growth. Oar wing is strictly a bone-joint deformation, whilst sticking-out feathers may occur temporarily on otherwise perfectly healthy limbs.

## Smooth-breasted Breeding Stock

In the smooth-breasted type, look for birds that have a profusion of long feathers and a fork or gap in the tail feathers: these will breed the best offspring and are heterozygous for the curled-feather gene (*Sb*). Such birds are easier to manage than curly Sebastopols, although there is the disadvantage of their breeding 25 per cent plain-feathered geese. When pairing the birds, some breeders use one of each type in the breeding pen and so breed both curly and smooth-breasted types.

## Colour Breeding

### White

Most Sebastopols are white, but many females have pale grey juvenile feathers on the rump around the preen gland. The females are usually greyer than the males at hatch - a useful auto-sexing characteristic in all white European geese (*see* Chapter 9). The occasional pale grey in the juvenile feathers does not usually matter because it will moult out by about sixteen weeks; however solid grey that remains in the male is a fault - this has become rather prevalent because buff geese have been crossed with white Sebastopols.

Sebastopols occasionally carry the Asiatic 'c' masking gene, from cross-breeding with Chinese in the past. Such goslings hatch bright yellow. Alternatively they may be rather smoky-coloured at hatch and seem to lack a spot (*sp*) gene.

### Buff and Blue

Buff Sebastopols have been produced in the UK and Belgium. If a pure solid pattern buff male is used with a white, then all the female progeny will inherit the buff gene and look mostly buff. Buff is sex-liked, and hemizygous in females. The F1 females will have white wing tips because they are heterozygous for spot (*sp*). The white Sebastopol female parent should be homozygous for *Sb* (curly) to retain this gene. All the F1 progeny will be smooth-breasted.

The F1 males will be heterozygous for buff, but will look mainly white with patchy grey; buff will not show. All-buff F2 Sebastopols and Buff Back Sebastopols can then be developed from these first crosses (F1) which are heterozygous for spot (*sp*) and solid pattern (*Sp+)*. Note that they are also heterozygous for dilution and not dilution (*Sd* and *sd+)* genes (*see* Chapter 7).

*F2 Buff-back smooth-breasted Sebastopols. In the first cross, the birds are heterozygous for the spot gene, and heterozygous for dilution. In the next generation, they can be pure for spot* (sp) *and be buff back; or they can be pure for no spot (solid pattern,* Sp+*). In practice* (sp) *is recessive and is difficult to breed out; Buff Sebastopols may have colour faults such as white flights and white under the chin for more than two generations in a breeding programme (*see also *Brecon Buffs, Chapter 14). Bart Poulmans*

*Buff Sebastopols. Bart Poulmans*

The spot gene needs to be eliminated for a Buff Sebastopol, or isolated for a Buff Back Sebastopol. Dilution also needs to be eliminated from all; Pilgrims (*Sd* and *Sp+*) also result from these crosses in the F2 generation if dilution is retained and spot eliminated. Note that this outcome presumes pure birds to start with.

Plain-feathered, buff birds crossed with a curled-feather Sebastopol (homozygous for *Sb*) will produce smooth-breasted feather types (*see* above), so the *Sb* gene must also be selected in the F1 and F2 generations. There are a lot of variables involved, so knowledge of the colour and plumage genetics is very useful in achieving an outcome; it avoids breeding numerous birds and aimless cross-breeding.

If the blue gene is wanted, then the sex of the blue parent does not matter because, unlike buff, the blue gene is not sex-linked.

## Eggs and Goslings

Sebastopols in Britain are largely kept as ornamental geese, but they can have excellent utility qualities.

Some are good layers, producing twenty-five to forty eggs per season, although they may not have such long reproductive lives as Brecons. The goslings are easy hatchers, fast growers, and readily put on weight.

Sebastopols can be good sitters, and being lightweight geese are not inclined to smash the eggs if they have a good nest. Despite their 'pantomime goose' appearance, they are good utility birds. As in all white breeds, the eggs are more hatchable in incubators and often beat the larger breeds' bigger eggs by 'pipping' earlier, usually between days twenty-eight and thirty.

*Auto-sexing white Sebastopols at two weeks: two females and one lighter grey male (left) in the spot pattern.*

## MANAGEMENT TIPS

* Sebastopols are relatively easy to keep clean if the turf is unbroken.
* If the ground is muddy, they do need paddling water. This is important in the breeding season, when a gander with dirty feet can make a mess of a goose when there is no bathing water for the birds whilst mating.
* Paddling or swimming water also keeps their feet in good condition.
* If the birds are to be exhibited, then it is essential to keep them in good condition all the time. Dirty Sebastopols cannot be cleaned up because the mud gets thoroughly ingrained in their feathers.
* They will also stay in better condition if they are not run with other geese, because the profuse feathers are very easy to grab in a disagreement.
* Protect birds from driving rain in cold weather: their plumage is not as efficient as normal feathering at repelling the elements.
* A tip from American breeders, who have superb specimens, is to feed some oats for good feathers.

# 11 Auto-Sexing Geese

## Introduction

Auto-sexing geese include the Pilgrim, Cotton Patch, Normandy, Shetland and the West of England; these are all descendants of the Greylag.

Telling domestic ducks from drakes is easy, but when it comes to goose and gander it is more difficult. Of course the birds always know, and for experienced keepers it is usually obvious too. But there has long been a fascination with goose breeds where the plumage colour is the instant giveaway. It is very useful to be able to tell the sex of a bird by its colour, even as a gosling, which must be why these breeds developed in the farmyard. A tradition of choosing a grey goose and a white gander has meant that auto-sexing breeds were produced by continued human selection from the goose colour genes. And the most fascinating part of the story is that the process is still happening in Australia today, where both Pilgrim and West of England females have even recently been found in a flock of white geese. The 'white' Pilgrim and WoE ganders are there too - but hiding!

## History of Auto-Sexing Geese

Nobody knows when sex-linked colour arose in geese domesticated from the Greylag stock, but it is certain that it pre-dated the development of the Victorian exhibitions and shows. There was a class for the 'Common Goose' in 1845, along with the Asiatic or knob geese and 'Any Other Variety', but after that point it seems that the farmyard goose of Britain did not find favour at the shows.

We know that there were white geese in Britain in the 1800s because they were used to improve the Embden. In Markham (1613) it was said that geese for breeding should be 'white or grey, all of one pair, for pyde are not so profitable and blacks are worse'. But the text is enigmatic, and it is not clear whether the geese should both be white, or both be grey, or one of each colour.

Written evidence that birds with sex-linked colour existed at all comes from Jenyns (1835), who does say 'the Common Gander after attaining a certain age is invariably white'. Harrison Weir (1902) also confirmed that:

> . . . the 'common' goose is a short and smaller bird than that of the 'farmyard' and 'better land' goose, being very plump and full on the breast, and seldom, when fully grown, exceeds ten to twelve pounds in weight . . . the ganders are invariably white, and that even if the geese are grey. But this may be and is probably attributable to centuries of selection as to colour, and is in some parts proverbial . . .

It is now generally accepted that geese with sex-linked colour were the Common Geese, before their place was usurped in the nineteenth century by the Embdens and Toulouse and their commercial crossbreeds. The useful goose which could fend for itself on the farm and on common land was gradually pushed out, and remained only in the rural backwaters of Britain, France, the USA and Australia, these far-flung birds all having arisen from a common European stock. They variously became known as the Pilgrim goose (where the female is dilute grey), and a group where the female is dilute grey but in a pied pattern, as in the Cotton Patch (USA), the West of England, the Shetland and the Normandy.

## The Pilgrim

### Pilgrim Geese in the USA

Hardly anyone had bothered to write much about the Common Goose until it was rediscovered by waterfowl breeder and geneticist Oscar Grow, in America. He coined the name 'Pilgrim' for a small farmyard goose that he developed as a breed in Iowa and named after the family's pilgrimage to Missouri during the depression of the 1930s. Paul Ives, who later became editor of the magazine *Cackle and Crow*, became interested in the birds after an article about them appeared in the magazine in 1934:

> Immediately I wrote to Mr Grow for more information; but except that tradition said they were very

common in New England in early days, that he had taken the ancestors of his flock from Vermont 40 years before (circa 1895), that they were said to have been a part of the Pilgrim Fathers' farm stock when they first reached Cape Cod and that there were very few in the country today, he had little or no information.

Grow felt that, prior to the development of genetics, people would not have been capable of understanding the inherited colour of the geese, so even if there had been 'some accidental instance of colour dimorphism in earlier times, the genetic significance of such a phenomenon could not have been understood and therefore would have soon been dissipated through aimless selection'.

Despite Grow's contention that the breed is therefore of comparatively recent origin, evidence has accumulated that auto-sexing birds were the common farmyard goose of America, and that their point of origin was Europe.

Hawes (1996) has determined that geese were probably *not* aboard the *Mayflower*. The first mention of any poultry was in 1623, when in a description of the village only 'divers hens' were mentioned, the stock presumably having arrived with the *Mayflower*. The second ship *Fortune* was criticized for bringing no provisions.

Barbara Soames (1980) does say that geese were mentioned on the bills of lading for the voyage: so maybe the geese were loaded, but did not survive that voyage. The geese must have come with later boats, however, because as well as Grow's foundation stock there are further examples in the USA. Peter Boswell (1841) referred to geese with sex-linked colour: ganders of 'fine white . . . the female either brown, ash or party-coloured'. Ives also found a small flock in Connecticut where they had been kept as a closed flock for thirty-five years, and Hawes cites a reference to a further flock of such geese in Alabama (*see* 'Pilgrim Breeding Stock', later in this chapter). Paul Ives finally concluded that:

> . . . from all the evidence I can find in a fairly complete reference library of old poultry literature which I have had fun in collecting for more than forty years, I have come to the rather fixed opinion that the ancestor of the Pilgrim was nothing more or less than the common or farmyard goose of early Britain.

## Pilgrim Geese in Britain

It has been said that Pilgrim geese were imported to the UK in the 1970s, but nobody that we have met can

*Pair of Pilgrim geese.*

recall this. Interest in the Pilgrim has grown however, and they were imported from North America in 1997.

During the 1980s, all the Pilgrim geese exhibited at the major events were owned by Ashton, Murton and the Domestic Fowl Trust. The birds originated, in part, from a pair discovered on the Fylde coast of Lancashire by Malcolm Stephenson in 1972. With the growing interest in waterfowl, more birds were discovered. There had always been auto-sexing West of England birds down in Devon and Somerset, and all-grey females doubtless turned up there too. Charles Martell of Gloucestershire developed a flock of silvery-grey Pilgrim geese; his original stock also bred typical grey females. And another flock of perfect grey females, but rather grey ganders, was found in Herefordshire, though the origin of those birds was unknown.

As Ives had also found, the conclusion is inescapable: the birds have been on the farm for centuries. To Appleyard they were the Common geese, and unremarkable. He was probably familiar with Harrison Weir's (1902) experience that 'the ganders are invariably white and that even if the geese are grey.' It is therefore to Oscar Grow's credit that he publicized information regarding this ancient breed, which could have disappeared without written recognition and breed status.

## Pilgrim Breeding Stock

Auto-sexing geese sound immediately useful and attractive. However, totally reliable stock is difficult to obtain in Britain, and in our experience they are the most difficult goose to get to breed true. There are two reasons for this: first, they have not been bred as exhibition geese until fairly recently, when interest in waterfowl

DESCRIPTION OF THE PILGRIM

**Carriage:** Just above the horizontal.

**Head:** Medium in size, oval. Bill medium in length, straight.

**Neck:** Medium length, moderately stout, slightly arched.

**Body:** Back moderately broad. Breast round, full, deep. Abdomen dual-lobed. Young birds and stock on grazing only are quite light and racy.

**Legs:** Shanks moderately short and stout.

**Plumage:** Tight.

**Colour of the male:** White with some traces of light grey in the back plumage, wings and tail. Out of a hatch, many ganders show pearly grey on the thigh coverts, rump and wings. Solid grey feathers should be avoided.

**Colour of the female:** Head light grey, front part white; the white extends around the eyes as a typical 'spectacle' pattern. White area grows with age up to about four years but should not cover all the head. Neck light grey, upper portion mixed with white in mature specimens. Back and sides of body light ashy grey, edged with white. Breast lighter grey, becoming paler, then white on the stern. Flights grey, secondaries darker. Tail grey and white.

**Bill, legs and feet:** Preferably orange (rather than pink).

**Eyes:** Blue in the gander, brown in the goose.

**Weight:** Originally low, 9-10lb (4-4.5kg) given by C. D. Gordon (1938), probably from underfed farm geese. Dressed weights of 8.8-9.9lb (4-4.5kg) are given for this type in Normandy, approximating to a live weight of about 14-15lb (6.3-6.8kg).

*American standard weights:* 14lb (6.3kg) male, 13lb (5.9kg) female.

*British standard weights:* 14-18lb (6.3-8.2kg) male, 12-16lb (5.4-7.3kg) female. Higher weights given due to better feeding and condition.

**Defects:** Completely white gander. White blaze on the breast of the goose; white flights. These features are evidence of cross-breeding.

**UK:** Standardized 1982. Status: Rare breed.

**USA:** Standardized 1939. Status: Critical (ALBC).

grew rapidly and note was taken of the American standard. Second, there is the colour dimorphism, so how can one be sure that a newly purchased 'white' gander is a Pilgrim? The truth is you cannot until you have bred from him with a guaranteed Pilgrim female and all of the progeny turn out correct, or nearly so.

Furthermore, even if all the male goslings turn out 'white', some will almost certainly carry varying degrees of pigmentation in the plumage. This is usually most evident in the secondary wing feathers, in the thigh coverts and on the rump. In some strains this grey will moult out during the second year; in others, as noted by Grow, the grey will intensify.

The female offspring should also be closely examined for colour. They often have white flight feathers, and as these increase in number, so a white blaze of feathers develops on the breast and the goose begins to become more like a saddleback. It is rare for a pair of untested Pilgrims to produce perfect females which are all grey apart from the distinctive markings on the face.

Photographic records of American and French 'Pilgrims' (illustrated in the *Magazine of Ducks and Geese*, Winter 1959) show females without white flights. The ganders vary a little in whiteness, several of them showing pearly grey on the thighs and wings. However, it is fair to assume that these photographs show selected breeding stock whose less-than-perfect siblings may have been culled. An indication of this is given by Gordon (1938), who, in describing sex-linked colour in geese, said that the 'females are mostly grey or parti-coloured with white areas on the head, breast [a fault] and the ventral regions'. Gordon was apparently unaware of Grow's new breed: these geese were in Alabama, and Hawes (1996) surmised that they could have been derived from Normandy geese brought first to Louisiana by French settlers.

A note from Delacour (1964) also confirms that foundation stock most likely came from France: 'The writer remembers seeing it [the Pilgrim goose] commonly reared on small French farms, particularly in Picardy, at the turn of the last century. It is now rather rare, having been supplanted by larger breeds.' A pair of Pilgrims is illustrated in *Plate 1. Domestic forms derived from the Greylag.*

### Pilgrim Goslings

Because the Pilgrim is a sex-linked colour breed, one expects the goslings to be auto-sexing too - but with newly hatched goslings this is not always as easy as it seems. If the individuals appear similar they may, of course, all be the same sex. However, look for colour differences in the beak, because newly hatched females have a dark grey bill whereas that of the males is a lighter colour. The males should be paler in fluff

*Pilgrim goslings. Unless the area is well keepered, fluffy goslings will be taken by magpies, crows, ravens and buzzards. The two darker grey female goslings have a dark bill. One has a slight colour fault on one wing, in that the grey fluff will become white primaries. The males are lighter yellow-grey in the fluff.*

colour, although they are often quite grey, but not as dark as the females.

Occasionally there is a paler gosling with a dark grey bill. This will turn out to be a white female which does not carry the correct Pilgrim colour genes.

As the goslings grow, beak colour eventually turns to orange. New fluff will begin to replace the first fluff after about ten days and the males will whiten. By four weeks, scapular feathers will begin to cover the wings and the white or grey will be obvious.

## Understanding Sex-Linked Dilution in the Pilgrim

The Pilgrim has the basic colour of the Greylag but in addition a mutation that dilutes the pigmentation. F. N. Jerome's work on colour genetics in geese from the 1950s gave a better understanding of these breeding characteristics:

> Since it was known that the single gene for sex-linked dilution [*Sd*] in combination with the solid pattern [*Sp+*] produced a light gray bird, similar to the Pilgrim female in color, it was realized that the presence of sex-linked genes could cause the difference in the color of the sexes. If one gene for dilution, as occurs in the Pilgrim female, would dilute the solid pattern from dark gray to light gray, it was logical to assume that the additive effect of two genes for dilution, as occur in the male, would further dilute the light gray to white.
> (Jerome, 1954)

In waterfowl it is the female, not the male, who determines the sex of the offspring. Females are heterogametic (ZW) whereas males are homogametic (ZZ). Humans are XX in the female and XY in the male - the opposite way round to birds.

The W chromosome (in the goose) is much shorter than a Z chromosome and carries reduced information - there is no dilution gene. Her equivalent Z chromosome does carry the gene, so the female is hemizygous for dilution and thus appears diluted grey. In contrast the male ZZ chromosomes each carry the dilution gene and make the male appear almost white. Thus 'white' in birds with dilution and no 'spot' are male only.

In contrast to the Pilgrim, white geese, such as the Embden and Roman, are white in both male and female because they have an extra gene. Instead of being solid pattern ([*Sp+*] as in the Greylag wild-colour) they have the recessive spot gene (*sp*). In conjunction with the dilution gene, this turns males and females 'white'. The combination of spot and dilution is best seen at hatching where white goslings - for example Embdens - show ghost grey-back markings in the fluff. Females are still darker than males because they have just one sex-linked dilution gene (Jerome, 1970) but then become white as they feather, albeit with a few grey feathers retained on the rump. In the Grey Back/Buff Back geese, the spot gene always shows its pattern because these birds lack dilution.

### Plumage Colour

*Colour Genes*

| | | |
|---|---|---|
| **'White' gander** | ***Sd/Sd*** | ***sp/sp*** |
| **'White goose'** | ***Sd/–*** | ***sp/sp*** |
| **Pilgrim gander** | ***Sd/Sd*** | ***Sp+/Sp+*** |
| **Pilgrim goose** | ***Sd/–*** | ***Sp+/Sp+*** |

| | |
|---|---|
| $sd^+$ | no dilution [+wild-colour] |
| ***Sd*** | **sex-linked dilution** |
| $Sp^+$ | solid pattern, no spot [+ wild-colour] |
| ***sp*** | **spot gene [saddleback]** |

*+ wild-colour geese are grey, like the Greylag.*
*Capitals denote dominance*

*NB Female gametes determine the sex in birds: the goose can only have one dilution gene. The male can have two such genes.*

The original research by Jerome suggested that the (*sp*) gene for spotting was sex-linked. More thorough investigation (reciprocal crosses) led him to revise this. The spot gene is believed to be autosomal and recessive to the solid colour (*Sp+*). Thus both male

*This 'Pilgrim' female was bred from a cross between a Pilgrim gander and a white Roman goose. She is heterozygous for spot and hemizygous for dilution. She does show 'spot' faults such as white wing tips, white under the chin and a white breast patch. Spot is variable in its expression, but is generally recessive.*

and female offspring of a spotted gander to a solid goose will be heterozygous and therefore unlikely to show the recessive spot. In our experience, however, it is not so simple. Dominance can vary. It can be affected by other genes and other processes within the cells. Expressivity itself is variable. Some heterozygotes can show elements of white (for example on some primary feathers, beneath the chin, and so on). Solid-pattern birds such as the Toulouse or Brecon Buff can have hidden spot genes that reveal themselves in future offspring in the way of the above faults. Getting rid of white spotting from the flocks can take many years.

DESCRIPTION OF THE WEST OF ENGLAND

**Carriage:** Above the horizontal.

**Head:** Neat, not coarse. Bill medium in length, straight, stout.

**Neck:** Medium length, moderately stout, slightly arched.

**Body:** Moderately broad. Breast round, full, deep. Back broad, slightly convex. Abdomen dual-lobed. Wings long.

**Legs:** Shanks moderately long.

**Plumage:** Tight.

**Colour of the male:** White with some traces of light grey possible in the back and rump plumage.

**Colour of the female:** Head grey and white, the grey extending down the neck. Even grey markings preferred to dispersed blotches. White feathers may be mixed with the grey. Forepart of head white, becoming more extensive in older birds which may then develop a white head. Thigh coverts grey, light-edged. The saddleback pattern is formed by grey scapulars, edged with white. Wing feathers (primary, secondary, tertials) are white.

**Bill, legs and feet:** Orange-pink.

**Eyes:** Blue in both sexes.

**Weights:** 16–20lb (7.3–9.1kg) male, 14–18lb (6.3–8.2kg) female.

**Defects of the WoE:** Undersized. Insufficient grey, or over-marking in the grey in the female.

**UK:** WoE Standardized 1999. Status: Rare breed.

**USA:** Cotton Patch status: Critical.

## West of England Geese and Cotton Patch

Whilst some farmyard geese in Britain were the pale grey, female Pilgrim colour, others were only partly grey in the females. 'In breeding the common domestic goose, some think the mixing of colours best, such as a white gander with dark grey geese, or at least grey and white or what is termed saddleback' (*see* Weir 1902). The female grey saddlebacks can still be found on farms in Devon and are kept by a few goose breeders who often start off with stock obtained direct from a farm. But beware: these non-selected birds may not be reliably auto-sexing.

There is now a written standard for the 'West of England': generally larger than the Pilgrim, the gander is white; the goose has grey feathers in a saddleback pattern, and some grey on the head and neck. Beak colour also distinguishes the sex of the birds at hatch. However, a logically consistent genotype is not yet available for this colour.

American Cotton Patch geese also exhibit this colour form. The ALBC considers that this breed was probably introduced from Europe during colonial times.

They differ from the WoE by having pink or orange-pink bills, lightweight bodies that are longer and sleek like a Greylag, and the ability to fly. These geese were used to weed cotton and cornfields up until the 1950s, and of course were named after the job they performed.

*Sketch of Sequanian female from Voitellier, 1918, indicating the darker head and saddleback pattern.*

## Shetland and Normandy Geese

Shetland geese are very similar in colour to the West of England. The Shetland is smaller, probably because of the hard living on these northern isles. The geese are described in detail in Bowie (1989) (*see* Bibliography). The Shetland is included in the 2008 British Waterfowl Standards and, along with the WoE, is listed as a rare breed. Similar geese are found in Normandy where the crested type is known as *huppé* (or *de Bavent).* In all cases of this auto-sexing colour, males are white; females are pied (in dilute grey) but tend to lose the grey colour on the head with age.

Even in 1924, Robinson (*Popular Breeds of Domestic Poultry - see* Bibliography) listed the Normandy goose amongst French breeds and stated that 'the Normandy geese resemble the Common goose of England and America before the introduction of the Embden and Toulouse. The females are grey and white; males mostly white.' Robinson's findings should not be surprising. Salmon and Perrier (1905) noted that in the Sequanian (common or commune) goose:

> . . . Sexual dimorphism is sometimes very pronounced, so that adult males are almost entirely white, but according to Cornevin, this has no fixed character. The Goose is common and wide spread as in the whole of Europe, its breeding is practised in all regions where there are grain, meadows and ponds . . .

Sequana was the goddess of the River Seine, particularly the springs at the source, to the north-west of Dijon. In addition, the Sequani were a tribe of Gallic people who inhabited the region including the Jura and Haute Saône in eastern France. However, Voitellier (1918) said the goose was most prevalent in the regions of the Loire, Normandy, Beauce and Champagne.

That the Sequanian type was indeed widespread is suggested by the geographically dispersed, but widespread, distribution of the WoE, Normandy goose and Shetland. In addition, such geese were also taken from

### SEX-LINKED COLOUR IN THE DILUTE SADDLEBACK PATTERN

White geese result from dilution (*Sd*) plus (*sp*).

However, white spotting (together with dilution) fails to do this in auto-sexing breeds such as the WoE, which retain the pied pattern in the females; the males are almost white. Cotton Patch geese USA have confirmed that spot does not govern auto-sexing. CP can be solid pattern (*Sp+/Sp+*), spot carriers (*Sp+/sp*), or homozygous for spot (*sp/sp*) They still auto-sex; females always have some grey plumage. This confirms that auto-sexing is in the quality of sex-linked dilution (*Sd#*).

*Crested Normandy geese in France, 2010. The males weigh 10–12lb (4.5–5.5kg) and the females 9–12lb (4–5.5kg). A range of 7.7–8.8lb (3.5–4kg) has also been quoted.*

Europe to the USA (to be called the Cotton Patch) and to Australia.

There has recently been a concerted effort in France to save the Normandy, which had become rare by the late 1900s. Originally the 'blue-eyed queen' of the smallholdings, this little goose has now become popular for exhibition. There were thirty-nine on show at the Exposition Nationale D'Aviculture at Tours in 2010, where they outclassed the grey Toulouse and the local white Touraine. With a programme for maintaining biodiversity, the breed's future now seems more assured.

*A pair of Shetland geese, UK; their weights are 10–14lb (4.5–6.3kg).*

## Auto-Sexing Geese in Australia

The Common goose also reached Australia. Andreas Stoll (1984) found auto-sexing geese in isolated parts of the continent, encountering them on numerous farms, and was able to trace one of these flocks back to an import from London in 1836. The birds varied from the typical Pilgrim to the heavier West of England. This suggests that auto-sexing birds were common farmyard geese at the time when the Embden was beginning to make its mark. As the white geese multiplied in both Britain and Australia, the obvious sex-linked colour was lost except in the outback (and in the west of Britain) where innovations on the remoter farms were fewer. Stoll's 'Settler geese' have now been formally recognized in the standards of Australia. They look like the American Pilgrim, but have been derived independently from the USA stock.

Out there to judge the geese at the Canberra Show in 2008, we were fascinated to find very good examples of these 'Pilgrim' geese from two breeders at the show, and to be told that flocks also existed elsewhere in Australia. What was even more interesting was our visit to another breeder of Embden geese, and there they were in that flock, too.

This flock occasionally produces lovely 'Pilgrim' and 'West of England' females. The comparable ganders are hiding: they just look white unless examined very closely. Ganders with modified dilution (*Sd#*) look white (especially if they are heterozygous (*Sd#/Sd*). They need to pass on just one dose of *Sd#* to a female offspring to reveal her grey plumage.

*Trevor Hunt's 'Pilgrim' (now called 'Settler') goose: Best Goose at Canberra Show, 2008.*

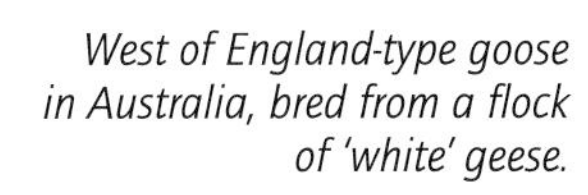

*West of England-type goose in Australia, bred from a flock of 'white' geese.*

*Pilgrim type bred from the same flock.*

# 12 Pied Geese

## Introduction

'Pied' is a term used to describe geese which have a heart-shaped mantle of coloured feathers on the back, grey thigh coverts, and a grey head. These markings are in contrast to the white body feathers. The pattern looks very attractive where it is symmetrical. This colour arrangement has also been described as 'saddleback' and 'spotted'.

Pied geese were described in Britain in the 1600s by Markham (*see* Chapter 1), and they were illustrated by the Dutch masters: Jan Weenix depicted a grey-back in *A monkey and a dog beside dead game and fruits* (1714, Rijkmuseum, Amsterdam), and even earlier, Gijsbert Gillisz d'Hondecoeter (1604–53) painted grey-backs in *Fowl on a river bank*, and Dirck Wyntrack indicated pied geese in the landscape (*Geese and Ducks at a Forest Mere*).

This colour pattern seems to have been popular in countries around the Baltic and North Sea. Not only are there grey-back Pomeranians, but small grey-back geese are found on Öland, an island in the Baltic off Sweden (*Avicultura International*, 1995). Also the Skånegås (Scanian goose) is figured as a 'spotted' or grey-back goose in a short article on colour in *The Magazine of Ducks and Geese* (USA 1956), and it now has a UK (2008) standard. In short, the 'type' and size of a saddleback goose varies depending on the location in which it is found. In north-east France and Belgium the Oie Flamande is bred in grey-back and grey. In Holland, the Twente is bred in pied and white; the grey Alsace breed (Elsässer) is now also available in pied.

*Oie Flamande in Belgium – Sauwens-Lambrighs.*

## Pomeranian

There is often a basic misunderstanding in Britain over the term 'Pomeranian'. This has arisen because these birds are often 'grey-back'. However, 'Pomeranian' is the area of origin, and it is a 'type', a particular breed of goose from north Germany and Poland, which comes in whole white, whole grey, and saddleback.

The grey-back Pomeranians are the most striking, and the best known internationally. The name is so popular that in the USA, the term 'Pomeranian' tends to be synonymous with grey and buff saddleback geese. The American standards do, however, describe the German phenotype.

### History

In Germany the Pomeranian is a specific breed which, according to German authors, has a long history. Geese have been kept for centuries in the marshy lowlands, originating in the Pommern (German) or Pomorze (Polish) region between the Oder and Vistula rivers (Szczecin to Gdańsk). Schmidt (1989) found, in a work called *Pommerania* (1550), that the geese of Rügen Island, just offshore in the Baltic (annexed to Pommerania in 1335), were driven to market on a regular basis. Geese were an important product of the region, and large numbers were bred for the delicacies of smoked goose breast, salted goose and 'goose lard'. These products were sold throughout the entire kingdom and the neighbouring countries, earning the goose the title of 'breadwinner of the farm'.

It is not known if the *Rügensche gänse* of the sixteenth century looked the same as the Pomeranian of today because there are few records. There has been the suggestion that Pomeranians were developed from an Embden/Toulouse cross. Brown (1929) quotes Moubray as saying that the saddleback geese 'are the production of white (or Embden) with the Grey (or Toulouse)', in his section on Pomeranian geese. However, Schmidt denies that the development of the Pomeranian had anything to do with such a cross. The type of the birds is completely different, the rather portly appearance of the Pomeranian having been arrived at by careful selection in order to build up meat and fat. Pomeranians are far more likely to have been developed independently from the Toulouse as the 'geese of the land'.

The Pomeranian was officially recognized as an independent breed in 1912, and in the following years two different but officially recognized types existed: the single-lobed and the dual-lobed (referring to the undercarriage). In 1929 the single, central lobe became an obligatory criterion for the breed. The oldest written description of the breed was attributed to Baldamus in 1897, where it was stressed that the frame was longer and higher than the other German geese, thus distinguishing it from Embden and Toulouse crosses.

*ABOVE: Pomeranian grey-back.*

### DESCRIPTION

**Carriage:** Nearly horizontal.

**Head:** Fairly broad, slightly flattened crown; stout beak; large, prominent eyes. Bill medium length, stout.

**Neck:** Medium length, stout, carried upright. Same thickness along its length.

**Body:** Plump with prominent, round breast; back long and slightly arched; single-lobed paunch, lobe centrally placed. Wings long, carried high.

**Plumage in the Grey-back:** Head and upper neck grey, lower neck white. Body mainly white; grey feathers of the thigh coverts and saddleback marking edged with white. Wings white.

There are also the **white** and the **grey** varieties.

**Bill, legs and webs:** Orange-red in the Pied.

**Eyes:** Blue or brown, depending on the amount of coloured feathers.

**Weight:** Males 18–24lb (8.2–10.9kg), females 16–20lb (7.3–9.1kg).

**Breed defects:** Dual-lobed paunch; keel on the breast; uneven markings in the pied.

**UK:** Standardized in 1982 as the Grey Back.

**USA:** Standardized in 1977 as the Saddleback Pomeranian (Grey and Buff).

**Status:** USA: Critical (ALBC).

*BELOW: Grey Pomeranian at the European Waterfowl Show 2002.*

*This group of grey-back (saddleback) geese, originating from German stock, show both dual- and single-lobed examples. C. Marler*

Schmidt also says that the breast and shoulders should be full so that the bird is 'egg-shaped' in long profile, emphasizing the breast. Between the legs, the central lobe must be straight and not misplaced to one side. The wings fit snugly to the body and the long back is slightly arched. Most important in the appearance of the gander is the neck; this must be strong, straight and of medium length, and of equal width along its length.

The head shape is also distinctive. The beak, viewed from the side, is deep. The culmen is straight and the crown of the head slightly flattened. The red-pink colour of the bill, and the orange-red feet indicate a typical East European origin (*see* Chapter 1). In its general demeanour the bird should look confident and arrogant, this being emphasized by the slightly flattened head, the convex topline of head and bill, and the prominent chest.

The distinctive head shape of the Pomeranian seems characteristic of the German breeds, the Steinbacher and German Embden also having this slightly convex outline of the crown and bill.

## Colour

Dürigen (1906) said that the colour of the feathers did not play a major part in the breeding process, but white was preferred. The grey type was grey all over, except for the lighter edging on the body feathers and the underparts, which fade on the stern to white. These birds have brown eyes, in contrast to the white birds which are blue-eyed.

On the saddlebacks, the grey on the head should extend a quarter to a half of the way down the neck, ending in a neat, symmetrical ring all the way round. Both this, and the heart-shaped saddle of feathers on the back, are difficult to obtain. Breeders of pied Pomeranians need to produce quite a number of birds every year in order to get a few with perfect colour markings. Some birds are over-marked and acquire the incorrect brown eye; others are under-marked and the coloured markings fail to close under the chin.

## Breeding Stock

It will be difficult to keep Pomeranians true to type in Britain because there have been few imports of the breed; most popular has been the grey-back. The birds tend to have rather orange feet and bill, rather than the 'hot orange' or red-pink that is required. Also, much of the stock, whilst having a good head type with deep bill and flat crown, is dual-lobed, and such birds should be disqualified as Pomeranians. This type of undercarriage is an inherited feature and can arise from parents which are both single-lobed, their type obscuring their genetic inheritance (genotype). In practice, these dual-lobed saddleback birds are being shown as Grey Backs, and the strong contrast between the white and grey plumage makes them very eye-catching.

### Temperament

Temperaments vary from strain to strain, but the Pomeranians we have had are not recommended for families with children. The females have a wonderful temperament, and the geese that we have hand-reared have been some of the most confident birds we have seen: they have fed from the hand, allowed nests to be searched while sitting, and enjoyed shows. Some of the ganders, however, have a temperament that matches their demeanour: they are alert and watchful at the gate, make a lot of noise, and put on a display of interest in visitors. But when people have backed away from them, or showed other signs of nervousness, these ganders have become the boss and a family nuisance; fortunately they can be retrained when moved to a different owner. Geese, and especially this breed, are very quick to pick up on human body language.

### Eggs and Goslings

German specifications are very thorough and give the weight of a Pomeranian egg as 7oz (170–180g). The breed is designed as a utility goose, but egg production does vary greatly. Our Pomeranians have been the first to start laying and have produced the earliest, easy-hatching goslings. They have also been the last to finish, sometimes having produced over forty eggs each.

## The Buff Back and the Grey Back

British saddlebacks result from crossing greys (or buffs) with white geese, and they resemble the Embden in type, not the Pomeranian. The Buff or Grey saddleback is quite a large, practical farm goose. Standard weights are similar to the Pomeranian, but well fed birds produced for the Christmas market, particularly those with an Embden in their ancestry, can reach over 22lb (10kg), even in adult females.

The 1982 British standard for these geese was derived from the USA 1977 Pomeranian standard, which correctly states that the Pomeranian paunch is single-lobed. In 1997, the dual-lobed Saddleback geese typical of the UK birds were distinguished from the true German Pomeranian.

*UK Buff Back geese.*

### Producing the Saddleback Pattern

Saddleback birds can be made from self-coloured birds (grey or buff) crossed with a white, such as an Embden. This is because the saddleback (pied) colour pattern is produced by the autosomal spot gene (*sp*). European white geese appear white because they carry both spot (*sp*) and sex-linked dilution (*Sd*), so by keeping the spot gene and breeding out dilution, saddleback geese are produced. It is best to use a white female because she is hemizygous for *Sd* (has just one copy of this sex-linked gene, *see* Chapter 7), and dilution must be bred out for the full expression of Grey Back and Buff Back.

If the buff gene is also required for Buff Backs (rather than the wild-colour for Grey Backs), then a buff male should be used. Buff is a sex-linked colour, so males are homozygous for buff and females are hemizygous. Buff males will breed females that show buff in the first generation, and phenotypically grey males which are heterozygous for recessive buff (*see also* Chapters 10 and 13).

# 13 Toulouse Geese

## Introduction

Toulouse geese have always fascinated the English. Managed on utility lines by the French, the exhibition birds have more status both here and in the USA where they were developed and exhibited as prestige geese. Today they are more likely to be treated as regal pets rather than as cramming machines for the production of pâté. It surprises me that this barbaric method of production is still permitted in the EU, where welfare standards for livestock ought to be better. It is time this practice ended.

## Breed History in France and the UK

Geese have been bred for centuries in the area of the River Garonne and around the regional centre of Toulouse. Even in 1555, Pierre Belon reported that there were two kinds of the goose: a large, heavy type, and a smaller, less profitable kind. Farmers had already selected a massive goose with tender meat for the table.

Harrison Weir attributed this breed to France, and quoted at length from a *Treatise on the Breeding and Fattening of Poultry* (1810). He commented that:

> . . . their distinctive mark is in having under the belly a lump of fat, which hangs on the ground the moment these birds walk. The fat is not very prominent until the month of October; it increases as the bird gets plump; it is called in the language of the country 'panouilhe'. Proceeding from Toulouse up towards Pau and Bayonne, this lump diminishes, the species become feeble and inferior, but in return they are better and more delicate when potted owing to the salt used . . .

Nolan illustrated the breed in 1850 using a print of the bird he had successfully shown, taken from *The Illustrated London News* (1845). He had obtained specimens from Lord Derby and was successful in breeding from them, distributing them to other breeders and winning first prize with them at the London Zoological Society's show in 1845. Nolan observed that some fine specimens of Toulouse:

> . . . have been recently introduced by the Earl of Derby, and are indiscriminately known as the Mediterranean, Pyrenean, or Toulouse goose, and from size and quality of flesh found a most valuable addition to our stock.
>
> With the exception of their great size, they resemble our common domestic geese, but of a much more mild and easy disposition: and what is more important to the farmer, they never pull the haystacks in a haggard. Their prevailing colour is blue-grey, marked with brown bars; the head, neck (as far as the beginning of the shoulders), and the back of the neck, as far as the shoulders, of a dark brown; the breast is slaty blue; the belly is white, as also the under-surface of the tail; the bill is orange-red and the feet flesh-coloured. The London Zoological Society have pronounced them to be unmixed descendants of the grey-lag.

Although the illustration provided by Nolan shows a fairly ordinary grey goose, Wingfield and Johnson (1853) commented on:

> . . . the unusual proportions of the abdominal pouch; this occurs a short period after they have emerged from the shell. Our own goslings have thus begun to assume this ordinary feature of mature birds when not ten days old; and at three months it will be seen almost touching the ground.

The same authors also said that Toulouse were not noted for going broody, and were good layers. A young goose which started to lay on 27 March (1853) continued to do so every other day until she had completed forty-five eggs.

Harrison Weir (1902) gives a fascinating account of how goose shows developed in Victorian times. Before the first exhibition of Toulouse in the gardens of the Zoological Society in Regents Park in 1845, he states that no geese were shown alive for prizes. This

*The Development of the Exhibition Toulouse Goose*

*Print from* The Illustrated London News *1845, sketch by Harrison Weir. These Toulouse were nothing like the exhibition bird of today. Nolan's bird scarcely had a keel or dewlap at the time when the 'Hong Hong goose' showed the latter feature.*

*Wingfield and Johnson (1853): the Toulouse is still just a grey goose.*

*Toulouse geese in France (Roullier-Arnout, 1880).*

*Simpson's line drawing in* The Feathered World, *1908. These Toulouse already show the full front and supports to the keel desired in UK/USA birds, the supports being considered a fault in Germany and France today.*

new style of poultry show (poultry alone), and the new breeds of geese such as Toulouse, Hong Kong Goose and Chinese, should have provided a much greater impetus for exhibition. However, the main show criterion – a relic of the time of the table bird – still remained the bird's weight.

The predominance of the Embden and Toulouse probably continued because the first *Poultry Standards* of 1865 gave descriptions of both of these breeds, but neglected the African, Chinese and Sebastopol, although these had been introduced to the UK by this date.

In his 1902 *The New Book of Poultry*, Lewis Wright recorded Miss Campain's account of her Toulouse:

> I started by buying a pair of his celebrated geese from Mr Fowler of Aylesbury who had then quite as good birds as anyone, if not the best at that time. The gander, I think was without exception the longest bird in every way I have seen and the goose was remarkably good in colour, very wide and deep, and not showing the least tinge of brown in plumage, but of a beautiful silver grey. This variety should be massive and heavy in appearance, the head should be broad and deep in the face, the beak being in a straight line from the top of the head to the tip, very strong and without any indenture or hollow in the top bill. The bill should be of a brown flesh-colour, the dew-lap should hang well down. The neck should be long and graceful. Both for the show pen and for breeding, the birds should be exceptionally well bowed in front, and 'keeled' deeply with their bodies almost touching the ground behind. In colour they should be rather dark grey on the head, neck, back and wings; rather lighter on the breast, gradually becoming lighter towards the belly, where it ends in pure good white.

Simpson's line drawing in *The Feathered World*, 1908, shows all the desirable features of the contemporary Toulouse. It was as described by Miss Campain, and looks remarkably like our modern UK exhibition Toulouse in that the keel appears to diverge on the breast to meet the dual-lobed paunch. This then fills in the hollows at the side of the keel, giving the bird a more full-fronted appearance.

During the first half of the twentieth century the Embden and Toulouse were the favourite exhibition geese, and were the only breeds which had a full, published Poultry Club standard until the 1954 edition. So although Roman and Chinese geese were advertised in the small adverts of *The Feathered World* in the early twentieth century, other breeds of geese were not often exhibited until the 1950s.

Geese have never merited many entries at shows. They are more difficult and expensive to transport than ducks; many birds went to the shows in London, such as the Crystal Palace, by rail. Only a few names appear in the show records and reports, and generally represent determined individuals who kept waterfowl for many years.

In the 1920s and 1930s, the Whitley line of Toulouse was frequently exhibited and photographed, and Reginald Appleyard also kept and exhibited the breed. But by the 1940s and 1950s, few geese were exhibited. The pure breeds of Africans and Toulouse became so depleted that by the 1970s an import of new blood was needed to revive several breeds. Christopher Marler imported Africans and Toulouse, plus Chinese, Embdens and American Buffs from the USA. Subsequent to the 1970s, several breeders also imported American, French and German stock.

## Exhibition Toulouse in North America

First imported into the USA in the early 1850s, the Toulouse was exhibited at the Albany County, New York Fair in 1856. The breed was not particularly popular at first, but Robinson (1924) considered it to be the most numerous of the standard breeds. It was most popular in the Centre West where there were large flocks of average quality, but the best

*American exhibition-style Toulouse in Holland, 2009 – Stijn Lemmens. Peter Jacobs*

DESCRIPTION OF EXHIBITION GREY TOULOUSE

**Carriage:** Almost horizontal. Birds move carefully to manage their weight distribution.

**Head:** Massive; effect exaggerated by loose cheek folds around the bill. Bill relatively short, deep; slightly convex top line. Uniform sweep from the point of the bill to the back of the skull.

**Neck:** Quite long, thick. A well developed dewlap extends in folds from the lower mandible down the neck.

**Body:** Long, broad and deep. Keel on breast is balanced by supports either side which diverge along the lower ribs, in front of the paunch. This gives fullness to the breast and keel when the bird is viewed from the front. Paunch full, dual-lobed. Wings large and strong, smoothly folded against the sides. Tail slightly elevated.

**Legs:** Quite short, stout. Hidden by thigh coverts.

**Plumage:** Soft and full.

**Colour:** Head and neck blue-grey, darker at the crown and on the back of the neck; back and thighs darker, each feather edged with white. Breast light grey, becoming paler under the body and finally white on the paunch and stern. Primaries and secondaries dark grey. Tail grey and white.

**Eyes:** Brown.

**Bill:** Orange or orange-red.

**Legs and webs:** Orange.

**Weight:** Males 26–30lb (12–13.6kg), females 20–24lb (9–11kg).

**Breed defects:** Rough wings, twisted keel, dropped tongue, white primary feathers.

**UK:** Standardized 1865, Grey; 1982, White; 1997, Buff, Rare breed (exhibition).

**USA:** Standardized 1874, Grey; 1977, Buff. Status: Watch (ALBC).

specimens, he said, were concentrated in the hands of breeders. Ives (1947) figured birds at the Government Experimental Farm (Ottawa, Canada) which had tremendous length in the body: these specimens already showed what is required of the American Exhibition Toulouse.

Left: *American Buff gander.* Centre: *First cross female.* Right: *Next generation Buff Toulouse – Peter van den Bunder*

Oscar Grow (1972) defined exactly what he wanted of the best. Not only was size, length and a well developed dewlap essential, the birds also had to display a well developed keel on the breast (the 'bow' in the rather nautical terminology once applied to the standards), which bifurcated as it approached the legs. From the front, this gives the appearance of a very broad keel, and when a bird stretches and flaps its wings, two broad supports are seen to run outwards on either side of the central keel to join the two lobes of the paunch. These structures fill in the hollow area seen in front of the legs of the rather different continental French and German Toulouse.

In addition to the Grey Toulouse, there are also a few White Toulouse (colour characteristics corresponding to the Embden) and Buff Toulouse (colour corresponding to the American Buff). The buff gene should be introduced into the Grey Toulouse by the male because the gene is sex-linked, and is hemizygous in the female. Thus American Buff males breed Buff females in the first generation. The sibling male F1 goslings look grey and are heterozygous for buff. A Grey Toulouse male with an American Buff female will breed grey females which have lost the buff gene. The male offspring are still heterozygous for hidden buff.

## Choice and Management of Stock

The Toulouse has remained the 'prestige' breed to show in Europe, Australia and the USA. A top quality

*Toulouse geese: the author's exhibition stock, 1998.*

bird will win Best Waterfowl: our best gander did it twice at the BWA Championship Show when it was held at the Three Counties Show Ground, Malvern, in the 1990s. Yet I am not sure why there is this obsession with the Toulouse; exhibition birds are huge, they do not breed freely, need special care, are expensive to buy, and are short-lived. They are not a beginner's breed.

*Buff Toulouse geese: the author's exhibition stock.*

## Selecting Toulouse

There are, in fact, many different qualities of 'Toulouse' ranging from ordinary dark grey geese (which are not Toulouse) up to the revered exhibition giants. As well as a continuous gradation between the extremes, there is a similar gradation in price and availability: Toulouse should be fit for purpose to avoid disappointment. Exhibition birds are expensive; pet quality birds are smaller and will live longer.

There is little point in breeding for exhibition purposes from anything other than the very best, because even exhibition parents can produce quite mediocre offspring. Birds related to American imported stock are usually far better growers than average stock, and will continue to increase in size well for two to three years. They peak in size, shape and carriage at about three years old, but for breeding, it is better to use a young, fit gander. Assess his potential by seeing the parent birds.

When selecting from a flock of youngsters, choose big birds with early development, for their age, of dewlap and keel. The keel should be straight and symmetrical, preferably broad at its base when the bird is viewed from the front. Also select birds with a large, strong head. The bill shape is important: it should have a straight or slightly convex topline, and not look slightly dished or long and 'snipey'.

### Breeding Stock and Fertility

Exhibition Toulouse geese have a reputation for being difficult to breed, and in general they are, for a number of reasons. First, it is essential to have a pair which suit each other and are not too inbred. Inbreeding leads eventually to a loss in fertility and viability. Fertility in the gander also depends on how fit and healthy he is; we have been able to breed from ganders which reached 30lb (14kg) in weight in the show season, but which lost weight to about 26–28lb (12–13kg) in the breeding season.

Birds should be given plenty of space and exercise to keep fit, and although ideally swimming water would be made available for mating, it does not seem to be essential in order for fit young ganders to be able to breed. It is, however, essential for heavier birds; besides which providing swimming water keeps all birds in good condition.

Pairs are much more successful than trios, especially if the gander is two years old or more. Older, heavy birds need peace and quiet, so that other birds do not interfere with them. However, a little confrontation through a fence, in a low-risk situation, can do wonders for the gander's ego and hormones.

### Diet and Digestive Tract Problems

Toulouse are not hardy grazers. Smaller geese such as Pilgrims and Brecons subsist well on correctly managed pasture with a small amount of supplementary food. Large domesticated animals have a higher food demand than their smaller, hardier cousins. They are large because of selection for rapid food conversion and weight gain. The development of dewlap, keel and full paunch – as well as the loose feathering – is a result of this purposeful selection. To reach their full potential, Toulouse must have access to short, tender grass and be fed waterfowl pellets in addition to their grazing. They are much keener to eat pellets than any other breed of geese, at all stages of growth, and as adults.

A heavy bird with a dewlap can suffer from dropped tongue. The dewlap tends to pull the skin of the lower mandible downwards, resulting in food collecting in the lower bill, and the tongue then dropping into the depression. This condition will need treatment by a vet (*see* Chapter 22).

Crop impaction and 'sour crop' is also more likely to happen with Toulouse than with other breeds, probably because of their shape. This occurs when insufficient grit is available for the bird, and fibrous material fails to pass through the gizzard. Such blocking of food passing through the system results in a hard lump collecting in front of the body cavity, in the proventriculus, and possible fungal infection (*see* Chapter 22).

### Care of Plumage and Eyes

Abundant, soft plumage also needs to be kept clean by regular bathing because the underparts, which touch the ground, can become muddy. Toulouse may need housing at night (even in a fox-proof pen) in periods of very wet, cold weather because their soft plumage will become permanently wet. Bathing water reduces the likelihood of wet feather (*see* Chapter 22). Buff Toulouse need even more care than greys because buff plumage is softer and more prone to 'weathering' than grey feathers.

Toulouse also suffer from foaming eyes more than other breeds. Ensure that their eyes stay clean. If bathing water is not available they must have a full, deep bucket of water for washing their face – but note the comment on drowning below.

### Temperament

Despite these drawbacks, Toulouse are popular pets, partly because of their *cachet*, and partly because of their personality. They are quite shy goslings in most strains and so should be handled and talked to regularly, and encouraged to hand-feed. If they are brought up in a gaggle of mixed breeds they tend to take a back seat when they are young, but gain more confidence as they become mature.

Young birds are quite nimble for their size and should be encouraged to walk and flap about with the other geese. However, by the time they are two years old, the exhibition birds will be much heavier and will need a quiet life on a piece of flat ground by the house. They are not very adventurous. The heavier birds have a more limited lifespan of perhaps up to seven years, compared with twenty years or more for an average goose.

### Eggs and Goslings

Toulouse are quite good layers. Exhibition geese, as well as smaller, nondescript strains, can produce over thirty eggs per year, between two and six years of age. The goose is not likely to go broody; if she does, do not encourage sitting because the eggs and goslings do far better under a goose or hen of lighter weight. The eggs, if fertile, are no more difficult to hatch than other coloured breeds; getting fertility is the more likely problem in breeding exhibition birds.

The goslings are rather small and slower on their feet than Africans and Chinese. They grow at a rapid rate, but do not overtake the Embdens, and it is worthwhile looking after them very carefully in order to maximize

*Toulouse goslings show their exhibition potential even at three to four weeks of age. They should have 'baggy pants' and a blocky body; a dewlap and keel must also develop early.*

their growth. Like the Embdens, brewers' yeast should be added to their diet. High-protein pellets, always preferred by the Toulouse, will be needed for a much longer growth period than medium or lightweight geese. Note that because of their shape, Toulouse are more likely to up-end in a bucket, get stuck and drown. Choose very safe water containers.

### Commercial Geese

Large, pure-bred Toulouse are of no use as commercial geese. The stock birds are expensive, they need concentrated food, and they need observing individually on a day-to-day basis. They are commercial geese in France because there is a much wider gene pool, and the birds are not the weight of the UK exhibition birds. Commercial birds are of the *sans bavette* variety. Also, white or buff feathers are generally preferred in UK table geese; grey does not seem to be a problem in France.

The most common commercial use of the Toulouse in the UK is as a cross-breed. A female utility Toulouse with an Embden gander is generally recommended to produce large, quick-growing birds without the danger of inbreeding. Fertility should come from the unrelated parents, and the higher egg production from the Toulouse. The cross breeds will vary in colour depending on the genes of the white gander, but crosses are frequently white with dilute grey feathers, often in a saddleback pattern by the F2 generation.

## French Toulouse

The French despair of our exhibition Toulouse. Speaking at the International Workshop on Domestic Waterfowl in Belgium in 2004, Pierre Delambre outlined why. The French, he said, saw the original Toulouse as a utility bird because in its area of production it was viewed almost like the smallholder's pig, where every bit of the animal was used and valued.

The volume of feathers and down was large because of the size of the bird, and as much as 1lb (0.5kg) could be produced. The enlarged liver produced the infamous *pâté*, and the fat was much prized for cooking as a substitute for butter and oil. And of

Left: *Exhibition Toulouse gander showing a dewlap and exaggerated cheeks; he has the desirable strong, deep bill. The other gander (below left) has finer features.*

course the flesh and fat finds its way into many traditional recipes, such as *cassoulet* and the various *confits* containing bits of duck and goose and available all over France.

It was only after the breed's importation into the UK and the USA 'when she almost changed her name to the *Pyrenean* goose' that the breed was developed for size. Indeed, Harrison Weir's illustration of the Toulouse at the 1845 Exhibition more closely fits the idea of the utility *sans bavette* Toulouse in France today: there is no dewlap, and the keel development is limited. It was in Britain and the USA that the type and size was exaggerated for exhibition, resulting in the 'hypertyping' referred to by Delambre, which he considered so detrimental for a breed.

The French still distinguish two sorts of Toulouse: *l'oie de Toulouse sans bavette*, meaning without dewlap, also known as *agricole*; and the dewlapped Toulouse (*à bavette* or improved type, '*industrielle*'). The *sans bavette* weighs 12.8–22lb (6–10kg). It is lighter in build than the exhibition Toulouse, has a more refined head, and also tighter feathers; it is more like the ordinary grey farmyard bird.

The French dewlapped Toulouse goose is slightly heavier at 17.6–26lb (8–12kg) and is quite big, reaching at least 33in (85cm) in height in the males. To obtain these dewlapped birds, heavier geese generally have to be selected, and also fed well. As the dewlap is selected, a keel on the breast and underbody also tends to develop, and plumage becomes softer and more abundant; this is especially so on the thighs, which are covered with long, strongly marked (edged with white) grey feathers. Delambre also points out that dewlap birds in France only lay between ten and thirty eggs, and a fertility rate of 50 per cent is considered very good. So what did he mean by 'hyper-typing'?

There is a tendency with exhibition animals and birds (especially those at the large and small extremes of the scale) to exaggerate size. Pushing size up or down – that is, away from the average – immediately imposes constraints: for example large geese find it difficult to mate, so fertility tends to drop.

When the shape is exaggerated as well, then features which come with very large, or very small size, also become exaggerated. In the case of the Toulouse, the fat of the deep, dual-lobed paunch becomes an encumbrance, as does the large keel which develops on the breast of the bird, and which runs under the body. Young birds manage their keel perfectly well as long as the land is flat, but they have difficulty as they age. The body actually touches the ground, and the weight out in front unbalances them. They also have a short life compared with more average geese, especially the females.

With their increase in weight and size, Toulouse also develop a very large dewlap under the chin. This is a loose fold of flesh which can descend as low as half way down the neck. It is often accompanied by folds of flesh at the side of the face and a heavy development of the crown so that exaggerated brows make the eyes lie too deep. These excessive trends do not belong to the French Toulouse, but have been developed in some exhibition Toulouse in the UK and USA.

## German Toulouse

Breeders on the continent believe that the German exhibition strain was developed by the Radetzky family from 1873. Adam Radetzky started a new line of exhibition quality Toulouse, and the family still owns world-class trophies won in 1900 in Paris, at the international exhibitions in 1911 in Turin and in 1912 in St Petersburg, and at the 1914 National in Berlin, where the Radetzkys' animals won the highest award of the show. A report by the son (Franz Radetzky, 1956) gives an insight into their racial history:

> In my youth little value has been placed on breast keels and head formation. Often, animals were given highest awards even with a crooked breast and keel, or pointed heads. English blood influenced the breed in a good direction.
>
> (Schmidt, 1989)

After the importation of English bloodlines of Toulouse, the keel became more prominent; birds developed bigger, well rounded heads, and the dewlaps were better. Already expensive in those days, 500 Mark (250 euro) was paid in 1925 for a couple of Toulouse from England (from the Watson strain). Radetsky's conclusion was that the English bloodlines gave more weight, big heads, a short deep beak, better necks and a better developed paunch. Breeders in Holland and Belgium regard these birds, which are low on the ground, to be more like the original Radetzky type.

Such birds are similar to those illustrated in the UK in the 1930s, which exhibited a prominent keel continuing straight to the paunch. However, there was no fullness in the breast, resulting in hollows on either side of the keel in front of the legs.

Exhibition birds in Germany are currently very different from those in the UK/USA. A more upright, longer-legged bird has been selected to lift the keel from the ground and to give more freedom of movement. This should also aid fertility. This shape is a real departure from the UK/USA Toulouse, and even from the exhibition birds on the continent earlier in the twentieth century.

*German-style Toulouse in Holland, 2003. Broekman, photo Peter Jacobs*

Toulouse continue to provoke discussion at all levels - as exhibition birds, as commercial geese in France, and as pets.

*Massive German Toulouse gander at Sinsheim, 2003.*

### MANAGEMENT TIPS

* Exhibition Toulouse are neither a smallholder's nor a beginner's breed.
* They are in demand as show birds and pets, but they need careful and individual care.
* Flat land, short grass and a pelleted diet are all essential.
* Avoid problems such as dropped tongue and impacted crop/gizzard by providing short grass, sand and grit.
* A shallow pool is useful for mating and will keep the birds' feathers and eyes in good condition.
* Toulouse are a placid, slow-moving breed. Don't expect to keep them with other lighter breeds. Give them their own space.
* Goslings must have a good diet of greens, crumbs, vitamin B and be exposed to daylight.
* They can be obsequious, quiet followers - don't tread on them by accident.
* Large adult birds have a limited lifespan.

*John Hall's Toulouse: Champion Waterfowl at the BWA National Show 2011. Photo - Rupert Stephenson and Tim Daniels*

# 14 Buff Breeds

## Introduction

Buff is not a popular colour in the wild. From the Pacific to the Atlantic, wild species of geese exist in shades of dark and silver grey, sometimes mingled with white. There are the unusual markings of the Bar-headed geese and a splash of colour in the Red-breasted, but goose plumage is generally conservative. Judging by the domesticated birds, there is a good reason for this. Buff feathers simply do not stand up to the rigours of a harsh climate, unlike grey or even white; they lose their sheen and are faded badly by spring, and can be damaged by sunlight even as they grow. This weakens the feather and spoils its water-proofing qualities, and this would prove a disadvantage in the wild and would select buff birds out of a wild population.

Buff is very much a domesticated colour. It probably arose as a mutation from grey and white farmyard crosses. Like the white before it, the colour caught the eye of breeders for both aesthetic and practical purposes, but it did not emerge as a breed colour until the 1930s.

The buff gene is quite hard to isolate, being both recessive and sex-linked. It may be that knowledge of genetics was needed for purposeful selection of breeding stock. The gene was eventually used to create the Brecon and American Buff, and the Celler in Germany. There are also 'clay-coloured' geese in Russia, and the gene lurks in the goose flocks of Australia, too.

As well as having the appeal of a new colour, buff has other advantages. The American Buff was developed as a commercial goose, and although this was not specified, it may have been because of the feather quality that buff was favoured. The feathers of buff geese pluck much more easily that whites and greys. The down is pale, and any pin feathers left behind on the carcass do not detract from marketing as much as on a grey bird.

Despite the obvious advantages of the buff, commercial growers stick with the white commercial Embden and Danish crosses. Hatchability and fecundity are probably more important, especially since plucking machines are so efficient. It is now left to the fanciers and the owners of small farm flocks to persist with the more unusual stains.

## The Brecon Buff

Although farmyard flocks had probably had buff 'sports' for decades before, Rhys Llewellyn of Swansea was the first to use these to standardize the Brecon Buff:

> In 1929, while motoring over the Brecon Beacons in Wales, I noticed a buff-coloured goose among a large flock of white and grey geese. This buff 'sport' interested me greatly and I purchased her there and then, with the definite intention of producing a breed of this attractive colour.
>
> (*The Feathered World*, 1934)

Unable to find a buff gander, he used a white, medium-sized Embden-type male. All the goslings from this first cross turned out grey, and he kept a gander from these. He fortunately acquired two other buff females from different hill farms, and using the young gander, which was heterozygous for buff, produced several buff goslings. Keeping the best buff gander, he then had an all-buff breeding pen; moreover he was able to avoid inbreeding too closely, because his original three females were unrelated. During the 1933 breeding season, the geese bred 70 per cent true to colour and type, and in 1934 he achieved 100 per cent true colour birds. Given the apparent rapidity of progress, one wonders if the 'white' gander was in fact a 'Pilgrim'-type male, lacking the (*sp*) gene.

Llewellyn's birds were deliberately reared in rather harsh conditions, as they were meant to be self-sufficient hill-farm geese. They had free range over a large area of good pastureland, and were kept in open grass runs at night. The eggs were hatched by the geese themselves; under these natural conditions the results were good throughout the breeding programme, even though the geese did not have access to swimming water. The birds were, not surprisingly, only fair layers, but their owner was pleased with their table

*Brecon Buff geese, 2008.*

*Brecon Buff.*

*American Buff. Pronounced feather partings on the neck of both breeds are characteristic of European geese derived from the Greylag.*

qualities and appearance, as they plucked out a good colour. Market requirements had changed in favour of a smaller goose, so he set out to produce a medium goose weighing on average 14lb (6kg) live. His standard allowed ganders up to 18lb (8kg) and females up to 16lb (7kg).

The Brecon Buff was exhibited at the Crystal Palace in 1933, where it caused a stir beside the standard white Embden and grey Toulouse. Llewellyn described the breed as not unlike the Embden in type, but of a lighter build, and more active. The buff colour was marked as in the Toulouse, the body feathers having a paler edge. The under-parts faded to white behind the shanks and under the tail. He liked a deep shade of buff throughout, and noted that the ganders were usually a shade lighter than the geese.

Although birds with an even buff did occur (Appleyard, 1933) and can still be found today, the marked sort is preferred. The actual shade of buff can vary according to exposure to sunlight, but even so, some individuals hold their colour better than others.

DESCRIPTION OF THE BRECON BUFF

**Carriage:** Slightly upright, alert and active.

**Head:** Neat; bill medium length, fairly deep at the base.

**Neck:** Medium length, no gullet.

**Body:** Plump, well rounded and compact. Breast full and round. Dual-lobed paunch. Strong wings, neatly folded to the body. Medium-length tail, carried nearly level.

**Legs:** Fairly short. Shanks strong.

**Colour:** Plumage a deep shade of buff throughout, with markings similar to the Toulouse. White fluff at the stern. White feathers may develop around the base of the bill with age.

**Eyes:** Brown.

**Bill, legs and webs:** Pink.

**Weight:** Ganders 16–20lb (7.3–9.1kg), geese 14–18lb (6.3–8.2kg).

**Breed defects:** White feathers under the chin, white primary feathers, orange hint to the pink of the bill, legs and webs, uneven paunch.

**UK:** Poultry Club Standard 1954; 1934, *Feathered World*.

**Status:** UK Rare breed.

A pinkish-brown is preferred to a tobacco colour – that is, almost a hint of blue rather than orange. Llewellyn standardized a pink bill, and pink or orange feet. In practice, the pink bills and feet go together, and pink is now the standard colour.

## Selecting Breeding Stock

### *Colour*

Pink-billed Brecons are not easy to breed. Stock that conformed to Llewellyn's standard was virtually unobtainable when we first wanted exhibition birds in 1982. Few geese were exhibited, and breeders were not particularly interested in the Brecon. Most birds advertised as 'Brecons' were quite orange in both bill and foot. The pink colour of the bill is difficult to isolate and even perfect parents can produce offspring with a great deal of colour variation ranging from pale rose-pink to salmon. Without selection to maintain the breed's colour points, even prior to the import of the American Buff, few good individuals were available.

The difficulty of getting reliable stock was compounded by the effect of nutrition and geology on bill colour. Free-draining soils derived from red sandstones, such as in Cheshire or Brecon, produce the best pink bills. An ideal stock bird bought in such an area and then grazed for two weeks on grass grown on solid clay, could soon change, the desirable pink bill colour rapidly becoming an undesirable bright orange. The reverse effect occurs when birds are fed only on wheat and pellets, or kept indoors, when the orange drains away because of the lack of the pigment provided by the grass. So whilst the birds look good for exhibition, their quality as breeders is unknown until they are grazed outdoors. Fortunately the popularity of Brecons over the last thirty years has meant that the colour of the birds is now much more uniform than it was, and quality breeding stock is easier to obtain.

The feather quality of a bird is much easier to assess. Obvious colour faults are white flights in the wing feathers and even a white blaze on the breast. As with the bill colour, continued selection has reduced the incidence of these faults. A patch of white feathers on the 'chin' below the lower mandible is the first sign of this problem (caused by the recessive spot gene), so it is probably better to leave such a bird out of the breeding pen unless it has other particularly desirable qualities. Even whole-buff birds can produce offspring with white flights.

Whilst breeders have sought the ideal pink bill, legs and webs, the buff feathers have tended to lighten a little in some strains. Care needs to be taken that the buff colour is maintained and does not fade to white too rapidly on the undercarriage. Incorrect colour tends to go with a pale eye instead of a deep brown one.

### *Type and Size*

Llewellyn described his birds as having a rather long, thin neck, but this is not accepted as typical today. The birds figured in 1934 were quite lean, and it seems likely that they were youngstock lacking the weight of well fed adults or exhibition birds. Grass-fed young birds are quite small and racy and will only reach the 14lb (6kg) average wanted by Llewellyn.

The Brecon was developed as a table bird and is ideally plump, with a round, full breast and dual-lobed paunch. Although its frame is not particularly large, so that it looks smaller than the Pomeranian, it carries a lot of weight. Grass-fed adult birds look just as round as breeders' birds fed on wheat and pellets, although there is a difference when they are picked up. Birds fed from a bag are much more solid and are deceptively heavy, often weighing up to 20lb (9kg) even though they look the same as their farmyard cousins. For this reason, the standard weights of the Brecon (in common with other geese) have now been given a range of values because the management regime plays a large part in the condition and weight of the birds.

## Temperament

Brecons have the reputation of being calm, friendly geese. This was particularly so of the first good original stock birds which we obtained from Lancashire and the Dolgellau area of west Wales. Hand-reared, the adults are usually trustworthy with children as long as they are treated well - though some strains, we found later, are less dependable. In the breeding season, ganders will be more aggressive and Brecons especially can become very territorial, in particular with what they see as competing pairs.

Brecons *can* be kept in sets of one young gander to four females, but as the birds get older they are more likely to pair preferentially, resulting in infertile eggs from the 'neglected' females. Brecons must be paired up well before the breeding season and preferably left in their pairs all year. The gander is very faithful to his female, once well paired, and partners cannot usually be switched around successfully.

## Eggs and Goslings

With all geese, I prefer to use mature females - two-year-olds rather than yearlings - as breeders: young birds lay small eggs, which produce small goslings, and so on. Brecons and Pilgrims are particularly prone to loss of size in the progeny if poor breeding and rearing practices are followed.

The breed is a good sitter. Grass-reared Brecons, fed

*Young American Buffs 2010 - Anne Terrell.*

as geese were intended, usually lay twelve to eighteen eggs in a first sitting, and then go broody. This was ideal for the farm without an incubator, where the goose did the job of rearing half a dozen goslings for the extra Christmas income. If these eggs from the first clutch are removed and incubated under broody hens, the goose will lay a second, smaller clutch, some weeks later.

Brecon and Pilgrim eggs are particularly difficult to hatch in still-air incubators in the UK. This is because of insufficient water loss, particularly early in the season, when egg shells are not very porous. Poor hatchability is probably due to the shell structure being different in these coloured breeds, as compared to white geese and Asiatic geese. To get good hatches of Brecons, it is necessary to regulate the humidity of the incubator very carefully, and if it is impossible to get the air sac large enough, the goose or broody hen is essential. In a very wet season, even eggs under the goose will fail.

## The American Buff

Buff birds were produced independently in the USA and named 'American Buff' to distinguish them from the Brecon. Although Schmidt (1989) does not seem to mention buff Pomeranians in his history of this breed in Germany, American authors consider that Pomeranian types were common in the USA in the first part of the twentieth century, and some of these must have been buffs.

Unfortunately, the development of the self-coloured American from Buff Saddleback Pomeranian stock was not documented. Even American authors such as Darrell Sheraw (personal communication), who have easier access to their own country's journals, have been unable to find much information.

The American Buff breeders, who developed and standardized the breed prior to its official recognition in the *American Standard of Perfection* in 1947, were primarily interested in it as a commercial goose. They did nothing to document its origins, or to advise others on breeding it for exhibition. One can therefore only assume that breeders would have used over-marked buff saddlebacks, which can show a great deal of variation in colour distribution, to produce eventually the buff goose. Alternatively, a solid pattern (*Sp+)* grey female could have been used to eliminate the *sp* gene.

Paul Ives thought that the American Buff was developed, through forty years of selective breeding in the USA, from the general farm goose found among the flocks of the peasant farmers of Pomerania. He considered that the pied buff goose or the Saddleback is a sub-variety of the American Buff. Certainly the first imports of American Buffs into Britain in the 1970s from a breeder in Kansas showed the reddish-pink feet common in East European birds, and John Hall (personal communication) recorded a tendency at that time for up to 10 per cent of the offspring to be well marked Buff-backs. The self-coloured American Buff, unlike its Pomeranian ancestors, has an even, dual-lobed paunch. This was a move away from the Pomeranian standard and perhaps quite deliberate, to make the new breed distinctive.

### DESCRIPTION OF THE AMERICAN BUFF

**Carriage:** Reasonably upright.

**Head:** Broad, oval. Stout bill of medium length. Any tendency for long, flat heads should be avoided.

**Neck:** Fairly upright and strong.

**Body:** Moderately long, broad, plump. Back medium length. Broad, deep, full breast.

Wings medium in size, and smoothly folded close to the body. Dual-lobed paunch which should follow through as symmetrical lobes at the rear.

**Legs:** Shanks stout, moderately long.

**Colour:** Feathers a rich shade of orange-buff, with markings similar to the Toulouse - that is, lighter edging on buff feathers, white fluff at the stern.

**Bill and legs:** Orange.

**Eyes:** Brown.

**Weight:** Ganders 22–28lb (10–12.7kg); geese 20–26lb (9.1–11.8kg). They should look significantly larger than Brecons, as the first imports did in the late 1970s.

**Breed Defects:** White feathers under the chin, white primary feathers, uneven paunch, dewlap.

**UK:** Standardized 1982.

**USA:** Standardized 1947. *Status:* Critical (ALBC).

### Selecting Breeding Stock

It is now recognized that American Buffs are larger than Brecons. Americans weighing up to 28lb (13kg) have been exhibited in the USA (*Fancy Fowl*) and birds bred by John Hall in Britain reach 26lb (12kg). The birds are 'rangier' (taller and longer-necked) than the

Brecons, and standard weights are now higher, to allow for the difference in size between the two breeds.

The American head tends to be a bit longer and with a flatter crown (rather like the Pomeranian), but an excessively long head and snipey bill should be guarded against when selecting breeders. The birds should be full in the body, and dual-lobed in the paunch, with this feature also symmetrical at the rear. Avoid birds with a dewlap and keel, which are Toulouse characteristics. Feather colour faults are the same as in Brecons with respect to white in the coloured plumage - that is, characteristics of the spot gene should be eliminated.

Paul Ives (1947) refers to the American Buff as an excellent layer, docile in temperament, and a fine sitter and mother: 'They dress off very nicely, and are a very satisfactory, all-round business goose.' As with all geese, behaviour will depend upon upbringing.

### Distinguishing Americans from Brecons

In the early 1980s, soon after the American Buffs had been imported, there was some misunderstanding over what constituted good Brecons and Americans, probably because there were very few good specimens around on show at the time. Brecon classes frequently contained orange-beaked, medium-weight birds. It was simply that, until the American Buff arrived, people had thought that any buff bird was a Brecon. After the import of the American strain there was a need to distinguish between the two. Probably due to an error, the top standard weight of the Brecon had been given as 1lb greater (at 19lb/8.5kg) than the American (at 18lb/8kg), proving a further confusing factor.

There were some good specimens, however. Fran Alsagoff's top quality American Buff goose, photographed in 1984 at the Federation Show at Stafford, clearly showed the superior size of her breed, compared to our pair of Brecons illustrated in the same book (Roberts 1986). Our Brecons at the time were only grass-fed and so were attributed a 'keel' rather than the dual-lobed paunch they now have. Americans from the Alsagoff strain and Brecons from that pair have gone on to provide the basic stock for many breeders since the 1980s.

## The Buff Gene

It is not known where and when the buff gene arose. Birds showing buff have been found in Australia, probably from UK stock; the American Buff was most likely developed from German Pomeranian stock in the USA; the Celler goose in Germany is basically a Buff Pomeranian. Even some Russian breeds have their buff variants (*see* Chapter 15).

*These Buff Tufted geese have been exported by Metzer Farms to Europe. Andréa Heesters and Peter Jacobs*

*German Celler goose at Hannover Show, 2009: this bird is particularly muted in colour.*

However, Buff geese standardized in Germany (the Celler) are not the same shade of buff as the Americans. Celler buff is more muted, rather as in the clay-coloured fighting geese from Russia; in contrast, American buff is bright and brassy. This distinction can also be seen in the muted shade of Buff Toulouse found in Germany, in contrast to Buff Toulouse developed in the USA. The shades may be different due to independent mutations - or they may look different due to selection of birds for depth of colour, affected by modifier genes. The sex-linked, recessive buff gene certainly arose in Europe or Russia, but where - and how many times - is a moot point.

The colour continues to fascinate breeders. There are now Buff Toulouse and Buff Sebastopols, and a medium-weight Tufted Buff Goose. Originating in the USA, it was developed by Mrs Ruth Brooks by crossing an American crested Roman goose with an American Buff (*see* Chapter 10 for the colour genetics of a buff × white bird).

# 15 Fighting Geese

## Introduction

Between the breeds from the Western Greylag geese and the pure Asiatic breeds produced from the wild Swan goose are the Russian breeds. These are not well known in Western Europe, but suggest, in the Tula and Arzamas, and also in the German Steinbacher, a fusing of Eastern and Western blood. But herein lies a problem, because distinct species are not supposed to interbreed and produce fertile offspring. However, both Chinese and African do produce viable crossbreeds with the European breeds, and in Russia such crosses have long been used in the selection of specialist breeds, as well as commercial strains.

## Russian Breeds Of Geese

The Russian breeds of fighting geese include the Tula, the Arzamas and the Gorky. The first Western reference to fighting geese may be Moubray (1842), who noted that:

> At St Petersburg . . . they have no cock-pits, but they have a goose pit! – where, in the spring, they fight ganders, trained to the sport, and so peck at each other's shoulders till they draw blood. These ganders have been sold as high as five hundred roubles each; and the sport prevails among the hemp-merchants. Strange that the vicious and inhuman curiosity of man can delight to arouse and stimulate the principles of enmity and cruelty, in these apparently peaceful and sociable birds.

The geese were said to have been bred in Russia from a very 'remote' period, and thus had been selected for 'sport' (Brown, 1906, quoting M Houdekow *Traveaux du Congress Internationale d'Agriculture 1899, St Petersburg, 1901*). They were also mentioned in Dürigen's 1906 *Poultry Book*: he, too, had seen such geese in Russia at the International Poultry Exhibition at St Petersburg in 1899 where, labelled 'fighting geese', they caused a sensation.

Two types were known, the Tula and the Arzamas. The Tula was mainly bred in the region south and east of Moscow. They were said to be generally grey in plumage, and some were also clay-coloured (buff); they weighed 13.5–17lb (6–7.7kg). They are described by Brown thus:

> Beginning with the nostrils the surface of the bill is ribbed, and the mouth, between the upper and lower mandibles has a rounded appearance, seeming even when closed to be partially open; the colour of the bill is pale yellow, with an ivory tip; the eye is large, full and nearly black as a rule.
> (Brown, 1906)

The heavier Arzamas was named after Arzamas, south of Gorky (Nizhny Novgorod). It is generally illustrated in white (van Gink 1931, 1941; Brown 1929); it is larger than the Tula at 17–22lb (7.7–10kg) (Brown 1929), also taller, and it has a longer neck.

The Gorky is smaller than the Tula. The Tula is more of a flat hovercraft shape; the Gorky has a more normal goose conformation but with a short, strong bill. Gorky geese also have a wonderful temperament, just like the Tula. They are very tame, laid-back birds.

Goose fighting still continues in Russia. In Loyl Stromberg's *Poultry of the World* there are amazing photos from the 1980s of this 'sport', banned in the early 1900s. Goose – or rather gander – owners hold and caress their birds before releasing a well matched pair of contestants to fight in the snowy arena in February and March. The birds fight in true Greylag style, using their wings to strike with the carpal spur. A ring of females also look on because, significantly, this is the breeding season when hormones run high and the ganders are pugnacious if roused. The birds illustrated by Stromberg are similar to the grey Gorky geese imported by Hans Ringnalda to Holland in 2003 – the last import before the EU closed the borders for poultry and waterfowl because of the bird flu threat from the East. Sigrid van Dort (personal communication) also notes:

> The Russians told me that there is no blood drawn any more these days (geese are too rare and expensive and important to the ego of

their owners when they win). Fighting geese are trained and they develop a callous on the shoulder where they hold each other. During the fight, the ganders will push each other within a drawn circle on the ice or snow, surrounded by spectators. In order to stir up passion, a goose is placed in the circle in a cage. The gander who wants to run away is the loser!

The fighting breeds are all built along the same lines: the body is very long, broad and deep with a full, round breast; the shoulder muscles are well developed, and the wings drop slightly. On handling, the birds feel heavy and very muscular. The neck is medium length but very strong. The head is very distinctive, characterized by strong cheek muscles, a wide forehead and a short, strong, deep bill which is convex on the culmen. The beak is bent down, like a hooked nose. The whole

*A group of adult Tula; a grey Gorky gander in the background.*

*Tula geese.*

*Tula geese in Holland. These are 'clay-coloured' (buff) rather than the normal wild-colour grey. Clay colour is recessive to grey. There are also piebald (with white). Hans Ringnalda/Sigrid van Dort*

*Gorky gander.*

*Tula goslings in grey, and 'clay-coloured' (buff) on the right.*

*The Gorky is smaller than the Tula. The Gorkys shown here are a buff and a grey.*

ARSAMAS (FIGHTING) GOOSE

Above and above right: *Arzamas goose, Edward Brown 1929 and van Gink 1931: 'The curved-bill one is no doubt the most interesting' – rather than the straight-billed and spoon-billed varieties.*

bird is built on powerful lines, with large wings, and legs set well apart for strength and balance.

In the *Animal Genetic Resources of the USSR* (FAO, 1989), the Arzamas is said to have been selected from local geese at Gorky, and was raised mainly on private plots. The limited distribution and low number of birds is because of its low egg production: eighteen to twenty per year. This is not surprising in view of the history of the Tula and Arzamas. Selection has been based upon muscle development, and a spirit and tenacity to fight which has been intensified by selective breeding, especially in the Tula. Numbers of the Arzamas declined from 3,700 in the 1974 census to 'few' in 1980. The Tula was not mentioned at all, indicating that it does not feature in commercial goose production.

The Kholmogorsk has not been used for fighting to the same extent as the Tula and Arzamas. It is larger, 20–25lb (9–11kg), and is very like a white African: it is therefore included in Chapter 8.

## The Steinbacher Fighting Goose

Many types of geese have originated between Eastern Europe and China as crosses between the 'geese of the land' and the Asiatic knob geese. These crosses, according to Dürigen, produced the fighting geese of Russia: the Tula and the Arzamas. The main difference between the two was the larger size of the Arzamas.

*A group of Steinbachers: a grey (background), two pure blues (left and right) and two heterozygous blues (foreground).*

*Head study of a Steinbacher showing the dark 'lipstick' serrations and dark bean. The older gander has a head shape reminiscent of the Tula.*

The Steinbacher Kampfgans was produced as a cross-breed in this same way as these Russian fighting geese, either as direct Asiatic/landrace cross, or through a cross with the Tula (de Bruin, 1995). This may have been deduced from the fact that trainloads of geese were brought from Russia to Thüringia before World War II. Ehrlein, a Steinbacher breeder from the former East Germany, also reported that crosses between knobbed geese and Pomeranian, Diepholz or Czech geese will produce first-generation progeny strongly reminiscent of the Steinbacher.

The fighting goose was established as a distinctive type in the Steinbach-Hallenburg area, and it was first properly classified in the Thüringer *Geflügelzuchter* in 1925. The breed was recognized and standardized in 1932 in its original grey colour, and then in the popular blue in 1951. Whatever its origin, the Steinbacher is a fusion of Eastern and Western blood, and is now a truly unique German breed.

### DESCRIPTION

**Carriage:** Slightly upright, proud stature.

**Head:** Neat; no knob or dewlap. Beak with straight or slightly convex culmen. Uniform sweep from the point of the bill to the top of the skull. Eyes large.

**Neck:** Medium length, strong, upright, straight.

**Body:** Stocky, with wide, full breast. Back strong and wide, slightly sloping. No obvious paunch in younger birds; any paunch preferably symmetrical and single-lobed. Wings long, carried close to the body. Tail short, carried level.

**Legs:** Thighs and shanks strong.

**Colour:** The Blue (Hellblau): Plumage light blue-grey; a slightly darker shade over the crown and down the back of the neck. Larger feathers show a sharp, not too wide, light-coloured edging. Underbody and back silver-blue. Stern white.

**Bill:** Bright orange with black bean and black serrations (a feature of the breed).

**Eyes:** Brown.

**Legs and webs:** Bright orange.

In the Grey, the markings and colour distribution are as the Blue.

**Weight:** Ganders 13–15lb (6–7kg), geese 11–13lb (5–6kg).

**Breed defects:** Sign of knob or dewlap (i.e. Chinese features). Lack of black serrations on the beak. Uneven paunch. Weak head/bill.

**UK:** Standardized 1997 Blue; 1999 Grey.

**USA:** Status: Critical (ALBC).

## The Steinbacher in the UK

The first UK mention of these birds may have been in BWA *Waterfowl* (spring 1987), when Hans Jensen of Denmark remarked that the Steinbacher was developed from a cross between ordinary grey geese and Chinese. The voice was a bit different to most other domestic geese, but not quite so noisy as the Chinese.

It was a rather good utility goose, laying twenty to thirty eggs, and is a good brooder and sitter.

Jensen also published the extant standard. The point of interest here is in the body shape, where he quotes: 'Body not too thin, keel [paunch] only just shown (double).' Since then, the German standards may have altered twice. A translation provided in the 1990s reads: 'Paunch in old birds acceptable.' This was followed in 2002 by: 'Abdomen: full and broad, well developed – possibly without paunch; however, small, single central paunch permitted in old geese. Bad Fault: one sided or double paunch.' The standard now asks for a typical German shape of a single-lobed paunch. In practice (just like the Pomeranians) the stock often breeds both shapes; the characteristic depends on both genes and feeding.

The breed was introduced into Britain in the 1980s and gained immediate popularity. It has a distinctive type and temperament, and was the only standardized blue goose. The birds have proved somewhat erratic in their breeding record, some individuals failing to breed at all, and others mostly reproducing only in small numbers. This has meant that even now there are still comparatively few in Britain.

## Fixing the Colour

Blue geese are known in flocks of cross-breeds and were referred to in an HMSO publication on geese. We have known 'blues' result from crossing white geese with buffs, but further breeding from the blues always resulted in a variety of markings, such a flecked and saddleback as well as self-coloured white, buff or grey.

The Steinbacher blue must be different because blues breed blues. The secret ingredient is probably from the Chinese. The Asiatic geese do not appear to carry the European saddleback (spot) gene, so if this is eliminated from the gene pool, then self-colour in blue can result. Chinas were crossed in Britain with buff geese and greylags by Frank Mosford of Clwyd, and attractive blue and buff 'Chinese' were produced as a result, even in the 1980s. The blue gene is likely to have originated in the Chinese gene pool, because blue Chinese have been known for some time in the UK.

Unlike 'blue' in ducks, Steinbacher blue looks blue in its homozygous (pure) form, where it is a delicate shade of pale blue (this is not caused by a lavender gene). In Blue Swedish ducks the blue is actually incompletely diluted black, and birds homozygous for blue often look 'white'; in ducks without the black gene, homozygous blue looks apricot-buff, and the colour is referred to as 'apricot'.

The original Steinbacher colour was, of course, grey and this is now standardized in the UK, too. But cross a grey with a *Hellblau* bird and all the heterozygotes are blue, in a slightly darker shade. Different shades of the F1 blue occur, and illustrate that it is probably the action of modifier genes which results in the specific blue shade.

Grey birds are less commonly available in the UK but in our experience are healthier than the blues, which can suffer from inbreeding in order to keep the attractive shade. Watch out for genetic defects such as reduced length of the flight (primaries), and wing problems similar to oar wing. Inbreeding will also lower egg production and fertility.

As well as the special blue plumage, the bill colour is also distinctive. Frequently people seem to think that the Steinbacher has 'teeth', because this appears to fit its fighting image. In fact all geese have these serrations for breaking off grass, but the 'teeth' or 'dentures' (a literal translation from the German) on most geese are usually the same colour as the rest of the bill. In the Steinbacher, the serrations are black and more noticeable. This does not mean that the birds are more aggressive, just that attention is drawn to the feature.

It is actually a breed fault if the black 'lipstick' is missing, and the birds really do look a bit 'undressed' without it. This also applies to the black bean at the end of the bill.

## Temperament

Although supposedly bred as a fighting goose, the Steinbacher breed in fact has a wonderful calm temperament as a pet. It is often the case that birds which are self-confident and aggressive with their own rivals are confident with people. On handling the birds, it is easy to feel how they have been selected: pick up a gander in the breeding season, and he will respond by tensing his muscles and arching his neck, quite unlike tame birds in other breeds. This confidence means that they are not frightened and therefore need not be aggressive with humans, who are not competitors.

With other geese, however, it may be a different matter. As long as the Steinbacher gander is recognized as the boss there is no problem: if everything defers to him, there is peace. But in the breeding season it will be different. Large Africans which have backed off earlier in the year will now no longer do so, and a determined fight will ensue if the birds are not properly penned. Pig netting which is generally satisfactory for separating pens of geese is inadequate because the ganders can get their necks through and fight. With most breeds this is usually a brief confrontation that ends in a quick separation, but the Steinbacher will hang on, and live up to its name as a fighting goose. Bloodied backs can result, and the situation becomes serious if the opponent is an African or Chinese, which,

*Steinbacher goslings are very responsive and appealing.*

when roused, will indicate by its tenacity where the Steinbacher has inherited its temperament.

In contrast, we find that Steinbacher ganders often get on together, and males that agree are particularly amicable, even in the breeding season. And just as the males are confident, so are the females: they are not retiring with humans or each other. However, a dominant female will not allow another female near her mate, so although the birds like to live in a flock, in the breeding season they may need to be paired.

## Eggs and Goslings

Although Jensen quotes twenty to thirty eggs per season and refers to 'a good utility goose', I should hesitate to describe them as such at the moment. They are not good layers, frequently laying only twelve to fifteen eggs per year. Note that the Russian fighting geese are said to be poor layers too, and this trait may have inadvertently been transferred to the Steinbacher.

A few astounding Steinbachers have laid up to forty eggs, for just a few years. But after that there has been a sharp decline, and by seven years of age there are no more eggs. Some birds lay just a few eggs in their first year, and then none at all. Utility geese should produce over thirty eggs annually and have a breeding life of twelve years or more.

The eggs are very hatchable in the incubator, usually requiring similar conditions to the Chinese eggs. The goslings are astoundingly beautiful and tame, and easily sexed by the size of their feet. The beak is completely black to start with, but changes to orange with the black bean and black lipstick as they mature. Ganders have very stout ankles by four weeks of age and have to be close-rung early.

Don't get Steinbachers if you want utility geese. The stock birds are expensive to buy, they are not heavy enough for the table (standard weights are 11–15lb/ 5–7kg), they may lay few eggs, and are in relatively short supply. People love them for their looks and temperament. First and foremost they are show birds and great garden pets.

# Part V
# Keeping Adult Geese

## Introduction

Geese which have been reared well are a pleasure to keep and need not be time-consuming to look after. However, if you like geese, and this is the main reason why many people keep these intelligent birds, it is fascinating to spend time with them to understand their movements, vocalizations and displays, which are still based on the habits of the Greylag goose.

With new stock always start off in a small way to see how the land behaves, and how the birds fit in with the family's daily and annual routine. Assess carefully how to cope in adverse conditions - drought, flood, heat-wave and also prolonged freezing conditions, when providing water may be difficult.

Plan how the birds will be housed, located and managed before buying stock. Livestock should be separated from traffic, protected from predators, and fenced off from young children.

In general, children and geese do not mix. Geese are very sensitive to human body language, and the inconsequential movements of young children are often viewed as a threat. Generally children must be over eight years of age before they develop an awareness of how their behaviour might affect others. A few bird-friendly individuals develop this skill early, but others may not.

In addition to the goose hazard for young people, there is also the hazard of water in pools and buckets: thus very young children must be fenced out of the livestock area for their own good.

# 16 Management of Adult Stock

## Space

There is no specific amount of land needed for a pair of geese: it depends on how much food is used to supplement their grass diet. The number which can be kept on a space such as a large garden will also depend upon the soil conditions (sand or clay), the water supply, the time of year, and the amount of rainfall. Free-draining soils create less mess than clay, which encourages water to lie on its surface. Basically, the accommodation should allow the geese freedom of movement to enable them to remain clean and fit.

The smaller the space available, the more important it is to shut the geese up at night, rather than leave them loose in a fox-proof pen. Goslings, in particular, produce a lot of droppings. If the birds are confined to a shed overnight, the muck can be removed from the premises and the grass stays cleaner. The actual quality of the grass is the most important factor for pet geese, but as a general rule, allocate a minimum of 33 × 66ft (10 × 20m) of well managed grass for a pair of garden geese (*see also* Chapters 6 and 21 for commercial figures).

*Weldmesh fencing, sunk into the ground, will exclude badgers. The two strands of electric wire also discourage most foxes, but they can clear-jump three feet (1m).*

## Fencing

### Stock Fencing

If the birds need to be confined, pig netting is the most durable material. Use the type with wire spaced at 6in (15cm) mesh, stapled to tanalized posts. However, breeders often use lighter-weight rabbit or poultry netting with a 1in (2.5cm) mesh to separate breeding pens of birds. This also keeps ducks in the right place and avoids the problem of ganders fighting by grabbing each other through the mesh.

In recent years badgers have become a problem, and strong weldmesh of gauge 12 (2.6mm) or 14 (2mm) is the best option. Although twice as expensive, it will last for many more years.

### Fox-Proof Fencing

Geese must be housed before dusk in order to avoid predation. Foxes and badgers can be a greater problem in urban areas where they are not afraid of people, and do not confine their activities until after dark. When protection against predation is critical, a high, 6ft (1.8m) fence is needed. This is best installed by a contractor, but do see examples of their work before giving specifications for an estimate.

Although chain-link fencing may seem to be the best material, it is expensive, and the holes are large enough for rats, mink and stoat to pass through; so the most commonly used material is rabbit mesh with 1in (31mm) hexagonal holes: 18-gauge (1.2mm) mesh in a 4ft (1.2m) roll is used for the bottom half, and lighter-weight 19-gauge (1mm) mesh for the top half of the fence. The two rolls are clipped together on to a middle straining wire, and the top roll clipped to the top strain-

*Geese and sheep graze up to the outside of this six-foot (1.8m) fence to keep the grass down. A fence in open ground minimizes maintenance (no hedge cutting) and maximizes breeding pens. The outside birds are driven in through a gate each night into sheds.*

*A single strand of electric wire on plastic insulators, close to the mesh, deters climbers such as the polecat. Two bottom strands of electric wire mounted on plastic insulators, supported on the posts, deter diggers and foxes. The grass will require maintenance by cutting or spraying so it does not touch the wire and shortcircuit the current.*

ing wire. The bottom roll has about 18in (45cm) buried under the turf, which is cut back and then rolled over the wire to anchor it securely. The wire is supported by 4in (11cm) thick posts sunk 24-35in (60–90cm) into the ground, spaced at 16ft (5m) intervals.

*A three-foot (1m) fence with electric wire looks better near the house - but a fox has been known to jump it despite the electric top strand.*

If sufficient mesh is available, it can be turned outwards as a 'floppy top' to deter climbers. However, it will collect wet snow and this will add a lot of weight to the fence. To deter climbing predators and foxes, which can run up straight fences, use a single strand of electric wire mounted on insulators around the top, instead of an overhang. To deter diggers such as badgers and foxes, especially in areas of light, sandy soils, strands of electric wire should also be used. These can either be mounted on insulators on the fence uprights, or on independent supports.

A mains-operated electricity supply should be used rather than a battery; it is more reliable and gives a stronger current. Minimal electricity is used in operating the system. The mains supply is connected to a transformer to deliver the correct voltage to the circuit.

## Housing

### Security from Predators

Overnight accommodation must be secure against polecat, mink, fox, stoat and badger. Vermin can squeeze through very small holes, and badgers can pull wooden sheds and wire mesh apart. In a badger-free

### TIPS FOR DESIGNING A FOX-PROOF PEN

**Maintenance:** This can be minimized by keeping the fence away from hedges which will grow into it and need clipping. There is the risk of damage to the fence itself if machinery is used close by.

**Maximizing breeding pens:** If the fence is set in open pasture, the grass on the other side of the pen becomes another grazing area for separating breeding stock.

**Gates:** These are weak points in the fence. If predators are a problem, have as few gates as possible. However, access through a gate out to pasture in the day and into the pen at night is useful for extra grazing.

**Size:** It may not be much more expensive to have a large pen built than quite a small one. Get estimates for two sizes if in doubt, but get the job done in one.

*The weldmesh of this gate is too coarse and mink and ferrets could squeeze through it, so it has been covered by netting. Gates and corners should be strengthened by extra posts and horizontal struts.*

area, wooden sheds are fine, but a brick outhouse with a solid door is safer if there are badger setts in the near locality. When rearing goslings with geese, the accommodation must be rat-proof and consist of a concrete floor or a new wooden shed – rats will soon get through old, rotting materials.

## Shed Requirements

**Size:** A pair of geese can be housed in a 4 × 4ft (1.3 × 1.3m) shed of the type designed as a dog house, allowing 8sq ft (⅔ sq m) floor space per bird; 10sq ft (1sq m) each is preferable for large birds. The shed needs a large door for easy cleaning.

**Design:** Geese do not require nesting boxes or perching bars, just a plain floor of exterior-grade plywood or tongue-and-groove boarding.

**Location:** The shed should not be located in a flood-prone position, and the entrance is best located in a corner so the birds do not run around the shed.

**Ventilation:** Good ventilation is essential. The window or opening should be covered with wire mesh, it should be high up (not at ground level), and face away from the prevailing wind.

**Purchase:** Garden centres and farm supply outlets often stock sheds for dogs, and these are often ideal for the geese. The *Yellow Pages* and newspaper classified adverts are also useful for information. If buying second-hand, consider that the shed may be infected with red mite if it has been used for poultry; avoid any potential infestation.

**Chemicals:** New sheds are treated with preservative, so allow a reasonable time to elapse after treatment for the fumes to escape. To repaint a shed with preservative, the stock must be re-housed elsewhere for at least a couple of weeks for the fumes to subside: they are an irritant to the throat and eyes, and stock should not be subjected to this. For this reason, only the exterior of the shed should be treated after purchase.

*Once they have been put in a new shed, geese will usually remember what it is for and where they are supposed to go. They are easily driven by using outstretched arms to guide them, and if they have been accustomed to being in a shed (rather than a fox-proof pen) they should drive easily. Always move slowly, talk to the birds, and use hand signals as well. Geese understand pointing, especially at food.*

### Bedding and Cleaning

Grass-fed geese produce more droppings than those fed concentrated food, but always provide bedding to absorb moisture so that the birds' feathers stay clean and waterproof. However, observe the following recommendations:

* Peat has been recommended as a bedding material in the past.
* The most commonly available material is a bale of white wood shavings.
* Rough-sawn waste from a timber yard (not fine waste from carpentry) is a good bedding material, though it must be preservative-free.
* Straw generates ammonia and aspergillae; hay is worse and should not be used.
* Avoid horse bedding materials which expand when wet (in case they are eaten).
* Sawdust and shavings will last for about a week, with topping up. After a week, maggots from manure flies will be growing and the litter is best removed. In winter, the bedding can be left to accumulate if it is dry.

One method to reduce work is to have a shed without a wooden floor, which rests directly on the earth. Litter placed on the soil will biodegrade, and excess moisture will drain away. Adding litter in small amounts to the top works on a free-draining soil, but not so well on clay. However, such a shed is not rat-proof and needs a weldmesh base which protrudes about 12in (30cm) beyond the edge of the house, to stop foxes or badgers digging in.

Discarded material, composted in a heap for some time, makes a good fertilizer. However, discarded muck should not be put back on to the goose pasture, to avoid a build-up of their parasites.

## Water

### Drinking Water

Geese are waterfowl and must have water freely available to wash and clean their nostrils and eyes. The minimum for a pair of geese is a two-gallon bucket of water per day in winter; this will need replenishing frequently on a hot day in summer.

Small geese (and goslings) can up-end and drown in a deep bucket, stuck head downwards, so choose a container of an appropriate size and shape - broad and relatively shallow, so a bird cannot get wedged head down, or on its back. The advantage of buckets is that the water is from a clean source daily and can also be used for feeding (see below). If it is topped up regularly, the geese will also dip and throw water over themselves to clean their feathers, and can keep themselves in reasonable condition without swimming water, though it is preferable for them to bathe.

### Swimming Water

On pasture, birds will stay clean without washing water. However, swimming water is more important for some breeds than others. On muddy ground, bathing is essential for Toulouse which otherwise become caked with mud underneath. Sebastopols must also wash their underparts. Female Sebastopols are a sorry sight in the breeding season when they have been 'trodden' (mated) by a muddy gander. Since the birds prefer to mate on the water, the females stay a lot cleaner if the gander's feet are clean. Water is preferred for mating but is only essential for large, heavy birds.

The birds do enjoy bathing water, but small pools quickly become fouled with droppings, and in a hot summer become smelly and a source of disease. Two useful types of portable pool are the sandpit bath (*see* Useful Contacts), and the Polypool; these both resist damage from sunlight and frost and can be emptied and moved so that one area does not become overused. Locate such pools near a hedge where the water infiltrates easily, or by a ditch; this avoids making the ground boggy and encouraging bacteria and coccidia.

A permanent pool needs a sump to drain waste water. A pool does not need to be deep to keep the geese happy: 6-12in (15-30cm) is plenty for washing and mating. The sides should be roughened and inclined so that goslings do not drown (very young birds should be denied access). The margin of the pool will become fouled with droppings so rounded gravel can be used to keep it cleaner. Droppings and mud wash down through deep gravel.

If there is a natural spring or stream, dams made of bricks and concrete, or railway sleepers, provide excellent pools; these need to be maintained because

*A sandpit bath for young geese, which jump in. It is not suitable for younger goslings which could get stuck and drown.*

*Swimming water is always appreciated and keeps birds in show condition. Activity on the pool in the stream also keeps it free of sediment.*

*Feeding wheat underwater saves wastage and fouling by wild birds. In hot spells of weather in the summer, ensure that excess wheat is not left underwater as it will start to ferment and go sour.*

they trap sediment. There are two great advantages to running water: it stays cleaner, and does not freeze in severe winter weather.

Swimming and paddling water helps to keep birds in top condition because they wash and preen more often and remove parasites more effectively. It also keeps their feet in good condition, avoiding calluses and bumblefoot (*see* Chapter 22).

## Food and Food Suppliers

### Grass

Geese are natural grazers: goslings know straightaway that green is good, and high quality grass is the cheapest way to keep and rear geese. It is not, however, a complete food for the year, and it must be kept less than 4in (10cm) long.

Its nutritional value varies over the year. The protein content is high in the spring and early summer, but declines rapidly after July, and remains low over the winter months (*see* Chapter 6). The high protein content correlates with the main growing season for the goslings. Adult geese on farm range, grazed with sheep and cattle, need little supplementary food over the spring and summer. Note that extensive range with other animals reduces the parasite load. However, geese on restricted range, where they are the only animals eating the grass, need wheat and pellets to stay healthy.

### Wheat

Wheat is a good basic food, higher in protein than barley and maize, though the protein content (9–12 per cent) varies with the variety and year of production. Barley can be used as cheaper fattening food. The whole grains are rather coarse and sharp so it must be fed with plenty of sand and grit available. Oats are said to be good for feather quality, especially in Sebastopols. The grains are rather thin and spiky and if rolled become floury. Yellow corn (maize) should not be used; it is a fattener.

If the grazing is good, feed wheat just at the end of the day. The geese should eat as much as they want over a twenty-minute period before they are shut up. If the grazing is poor or dirty, then more wheat should be fed, as long as the birds do not get too fat. This is best done by putting the daily amount in a bucket of water. The wheat is softened by soaking and is then milled up better in the gizzard, and wild birds cannot get at the food. Adult geese may eat up to 7oz (200g) per head of wheat in the winter, but less in the summer with the warmth and the greater amount of grass.

There is no need to feed the geese in the shed overnight. They should not be left dry food without water, and if they are left a bucket in the shed, they will foul the bedding.

Wheat stores well. It is best bought by the bag from a farm supply outlet, or direct from a farm. In contrast, pellets are manufactured and have a shorter shelf life. Pellets must be purchased on a regular basis, stored in a cool dry place, and used before the date of expiry of the vitamins.

In general, geese prefer wheat to pellets, but Toulouse like the manufactured food and keep weight on much better with a pellet ration. Unlike Pilgrims, they have not been bred for a hard life, and are not such good food converters. In addition to wheat underwater, give them a wheat/pellet dry mixture each morning and evening.

### Layers' Pellets

These pellets may be manufactured by a local food mill; use the *Yellow Pages* (Animal Feed section) to find a supplier. However, they often contain undesirable additives. Egg-yellow colouring is added to hen-laying rations and it is preferable to switch to a breeder ration, sourced from a specialist in the spring. Also check the type of preservative used: products such as ethoxyquin, fed over the potentially long lifespan of a goose, may be harmful. Some companies, however, do not use preservatives (*see* Chapter 21).

Layers' pellets contain more calcium than growers' pellets. A high production hen-layer ration can be 4 per cent calcium. Calcium for eggshell formation is absorbed from both the diet and medullary bone (bone that acts as a source of rapid calcium mobilization for eggshell formation).

Note that a growers' ration has less calcium. The calcium:phosphorus ratio is 10:1 for laying birds, whereas for growing birds it is 1.5–2:1. Excess calcium causes problems for growers, affecting the availability of other essential minerals.

## Breeders' Pellets

These are available in spring from specialists who also manufacture products for rearing waterfowl (*see* Chapter 21); check with their head office that there is a local distributor. Breeders' pellets should have high quality ingredients, giving sufficient amounts of the vitamins A, D, E and K and trace minerals (selenium, copper, zinc and manganese) for a viable embryo. Higher protein levels also ensure protein quality in the egg, and there should be a correct balance of omega 3 and omega 6 fatty acids. Essential amino acids must be present in the correct proportion, so fishmeal is sometimes added to supply these.

However, geese often prefer quality spring grass to pellets, and grass can provide for most of their needs if just one small clutch of eggs is wanted. Wild geese obtain all their nutrient demand, including protein and calcium, from grasses, sedges and roots – but of course they have extensive choice. They also lay only a few eggs, and fail to breed at all if the food supply is poor.

### SOURCES OF NUTRIENTS FOR VIABLE EGGS AND GOSLINGS

* **Vitamin A:** Green feeds, yellow vegetables. The vitamin is easily destroyed by light and heat. A lack of this vitamin is a common cause of poor hatchability, weak goslings and poor disease resistance.
* **Vitamin D:** Sunlight. Goslings kept indoors will need supplements. Deficiency causes weak bones that will bend and deform.
* **Vitamin E:** Seed germs – as wheat. A deficiency results in poor hatches of weak goslings.
* **Vitamin K:** Green feed.

***Comparison of a Breeders' Ration with a Layers' Ration***

| | *Breeder* | *Poultry Layer* |
|---|---|---|
| Oil | 4.0% | 3.5% |
| Protein | 17.0 % | 16.00% |
| Fibre | 4.0% | 7.00% |
| Ash | 10.0% | 15.00% |
| Methionine | 0.36% | 0.35% |
| Moisture | 13.8% | 13.80% |
| Vitamin A retinol | 15000 iu/kg | 8000 iu/kg |
| Vitamin D3 | 45000 iu/kg | 2000 iu/kg |
| Vitamin E | 150 iu/kg | 10 iu/kg |
| Sodium selenite-selenium | 0.45 mg/kg | not specified |
| Copper sulphate-copper | 20 mg/kg | not specified |
| Main ingredients in descending order | Wheat, soya bean (heat treated), wheatfeed, distiller's dark grains | Wheat, wheatfeed, sunflower extract, maize gluten feed, barley, rape seed extract |
| Other ingredients | calcium carbonate, mono calcium phosphate, salt | calcium carbonate, di calcium phosphate, salt |

## Sand, Grit and Calcium

Sand and grit are needed all year round so that birds can make the best use of their food (*see* Chapter 6). During the breeding season, the goose must have access to mixed poultry grit containing flint, oyster shell and limestone chips. Flint is the insoluble, hard grit which grinds up wheat in the gizzard; it is made of silica ($SiO_2$), and although the crushed material can have very sharp edges, this does not seem to adversely affect the gizzard.

In spring the goose needs extra calcium, over and above the normal amount supplied from grazing, for eggshell formation. Oyster shell and limestone supply calcium as calcium carbonate. Supplementary calcium can also be supplied as calcified seaweed (*see* Chapter 6) and as baked eggshells (which will also supply phosphorus). In the wild, Pinkfoots nibble old eggshells when they return to nesting sites. Do not, however, use eggshells from elsewhere because this could import disease.

Extra calcium is said to provide a thick shell, making

*Feed a mixture of wheat and pellets dry, in a bowl. Pellets disintegrate in water, and go mouldy if damp.*

it difficult for the embryo to hatch. However, without it the goose will suffer from soft-shelled eggs and oviduct problems. Females must be allowed to help themselves to what they need. Calcified seaweed also provides other trace elements (magnesium, manganese, copper) to balance the extra calcium.

### Food Containers

Food is best put in a clean container rather than thrown on the ground, as there is then less wastage. Look for a non-tip receptacle which is heavy and has a broad base. Old cast-iron casseroles, and saucepans with the handles removed, are useful: never throw away old kitchenware which can be used for feeding and rearing. Galvanized troughs are a good design for larger numbers of birds, and wide, low buckets. Keep pelleted food out of the rain; it becomes soggy and unpalatable when wet, and will quickly go mouldy. It is best fed for a short time at the end of the day, which also avoids fouling by wild birds.

## New Birds: Transport and Arrival

Before collecting new birds, consider carefully the containers you will use for transporting them. A large, strong cardboard box (of the kind used for a television) with several holes (3 × 2in/8 × 4cm) cut out for ventilation may seem ideal. However, bear in mind the time of year and use your common sense. Many waterfowl are bought in summer, as the goslings come up to adult feathers, but this is the worst time of year for transporting them because of the heat. Goose-down is designed for winter, so imagine what it would be like sitting in a well insulated cardboard box, under a goose-down duvet. Geese overheat rapidly in these conditions, and can die during even a short journey.

### CHECKLIST FOR TRAVEL

* The birds must be securely restrained for transport.
* Find stout cardboard boxes, ideally one for each bird (so they do not overheat), and sufficiently large for the bird to stand up and turn round.
* Make ventilation holes by slashing triangular holes with a knife, and remove the cardboard. Weave the top flaps so that it is secure and then tie with string. If hot, fix a piece of chicken wire across the top as a lid instead.
* Do not stack cardboard boxes as they can collapse.
* A plastic poultry crate or a dog cage of the appropriate size (it must be taller for geese) can be used for transporting birds in numbers, as it is easy to clean and disinfect. It is also airy, much better in hot weather, and water containers can be attached.
* Good footing should be provided. Use newspaper followed by a layer of straw (for a good grip), then shavings.
* Boxes should be placed on the back seat of a car and not in the boot unless the back seat is folded down. A sealed boot must not be used. Ensure there is sufficient ventilation by opening windows or using air conditioning.
* The vehicle and boxes must be kept cool when loading; park in the shade even on 'cool' days. Never put the birds into the vehicle until you are due to go.
* If a trailer is used for transport, make sure there is adequate ventilation for the birds both when travelling and when the trailer is stationary.
* Food and water must be provided for journeys over twelve hours.
* Animal transport certificates are not required if the birds are kept purely on a hobby basis – that is, where there is no commercial interest. If birds are transported on a business basis then an ATC is required, and if the ducks and geese are carried for sale on a journey of 40 miles (65km) or more, the transporter must hold a certificate of competence.

In many ways, new geese are easier to manage than new chickens or ducks, but more difficult in other ways. They are more intelligent and need sympathetic handling. A tame pair of geese will be no problem for their new owner. Placed straight into their new shed, they can be let out after ten minutes. Most birds, except for flighty Chinese, will not run off, and they will remember the shed when driven back to it later in the evening. After a couple of nights of being driven into the shed, this becomes a simple routine.

## Catching and Handling Birds

Never run birds around or use nets when catching geese. Nets have to be used at large establishments, but there is no easier way of making birds nervous. If the birds are tame, simply pick them up. Otherwise, drive them to their usual shed and pick them up in a confined space. Geese must not be suspended by the legs, in the manner that chickens are sometimes (inadvisably) carried. Waterfowl legs are not as strong, and the bird can be injured - and anyway it is illegal to handle birds like this. It may be necessary to catch hold of the bird by the wings or neck, but when carried, it should be held facing backwards, and its weight should always be supported by your hand and arm.

Thus if you are right-handed, hold the bird with its body on your right arm so its neck is under your arm and its head looking behind you. The right hand should hold both legs, parted, and the left arm can be placed across its back, or even hold the flights of both wings together if the bird tries to flap. Carrying a bird by the wings can result in dropped wings because this strains the joints badly.

*When holding a goose, support the bodyweight and tuck the head and neck under your arm. The right hand is holding the bird's right thigh, so the legs are not squeezed together. This female has lost feathers from her head during mating.*

*Clip five to seven outer flights on one wing only. Clip before the coverts, and do not expose the hollow quill to dirt and bacteria.*

## Clipping

If new birds seem nervous and capable of flight (in the smaller breeds), then clip five flight feathers off just one wing. Using a pair of scissors, snip the five flights that lie just before the coverts - that is, less than half way along the length. Do not expose the hollow quill. If the birds do try to fly, the unbalanced wings will put them off. Birds need only be clipped once a year, after the new flights are fully formed and all the blood has been withdrawn from the new quills (*see* Chapters 5 and 20).

Geese which are accustomed to flying around can manage to fly away quite easily, especially if they have an extensive range to forage and are light. However, most domestics will not bother unless they are frightened. Risky situations are locations on a hill with lift into a good wind, and also broody geese that have lost weight.

## Worming

Geese can carry a number of parasites, the most harmful of which is the gizzard worm. All geese should be wormed on transfer from one place to another; if this were done routinely, it would lower the incidence of this pest. When acquiring new stock, always ask if and when the birds have been wormed, and do this yourself if necessary. The parasite proliferates by eggs being passed out with the droppings and being picked up from the pasture. This is why litter from the sheds should be used elsewhere and why geese raised entirely on grass need extensive, clean grazing.

Worming should be done strategically - that is, the use of anthelmintics should be limited to key times and

*Worming a goose: hold the crown of the head in the left hand and open the beak. Use the syringe (no needle) to hold the beak open, and introduce the liquid wormer to the back of the mouth. It is essential the goose swallows so that liquid does not enter the airways.*

vulnerable individuals, so that parasite resistance does not build up through over-use (*see* Chapter 22).

## Hazards

Items that are hazardous to geese include pieces of aluminium and glass, nails, lead shot and staples. Geese are inquisitive creatures, and will pick up any bits of interesting hard material for the gizzard – in the absence of grit they will seek out anything hard including dangerous substances which can be ingested, sometimes with fatal consequences. Lead is particularly dangerous when it is finely ground up in the gizzard (*see* Chapter 22). Always collect any discarded foreign material before the geese do; worms and moles are constantly turning the soil over, exposing new debris left from many years ago. If vermin is shot on the premises, use bismuth shot instead of lead.

### String

Binder twine is the farmer's friend and much used, but always use it sparingly, fix it well, and leave short ends. A goose will spend hours twiddling at something interesting, and can swallow string. If this happens, a short piece (if you know the length) can be removed by pulling gently, but only after a lubricant – such as cooking oil – has been applied down the throat. Sit and hold the bird and the twine while the cooking oil has time to go down the throat and lubricate it, before you try extracting the string. If a large amount of string has been swallowed, more harm than good may be done in trying to pull it back through the gizzard and gut – it is best to avoid this occurrence altogether by being tidy. Wire and large plastic electrical ties are safer alternatives to string.

### Vehicles

Rotating wheels really seem to annoy geese, particularly ganders in the breeding season. This does not matter if it is the wheelbarrow, but it is quite common for birds to be killed in the farmyard, crushed when attacking a vehicle. All birds are best kept separate from traffic.

### Toxic Plants

Poisonous plants include ragwort, foxglove, hemlock and nightshade. Adult geese generally stick to nutritious grass, but hungry adults and goslings are at risk. Don't allow access to these weeds, and remove small quantities by hand if necessary. Ragwort is still dangerous when sprayed, and if uprooted, care must be taken to remove all the plant otherwise it will regrow. A number of garden plants are also toxic: laburnum, wisteria and daffodils, for example. Geese may not touch these, but it is best to avoid planting new hazardous materials.

### Heatwave

When deciding on the area where you will keep the birds, consider how you will provide them with shelter from bitter winds in winter and hot sun in summer. If there is no shelter at all, then plant fast-growing evergreens that will not be eaten by the geese, and which can also be used in the future for nest screens. Extra deciduous trees for summer shade are also desirable: these could be fruit trees, though beware of large apples dropping off the trees and hitting the geese on the head. Geese do like to eat a few windfalls, but do not give them fermented apple waste. Young plants and saplings will need protection because the geese will ring-bark them.

## Gales

Locate housing and fox-proof fencing where it is safe from falling trees. Foxes have been known to enter compounds when trees have brought down fencing.

## Prolonged Freezing

If the birds live out in a fox-proof pen, they will need shelter and ad-lib food during any prolonged spell of freezing weather when temperatures stay well below zero. Bales of straw as a windbreak and straw on the ground can be used, though the birds may choose to ignore these. They will sit on their well insulated breast and tuck their feet securely into their flank feathers. Their legs are adapted to withstand cold temperatures, unlike tropical species which get frostbite.

Geese suffer less in these conditions than in a heatwave, especially if they have access to running water which does not readily freeze because it is fed by groundwater at a steady temperature. However, remember to empty all buckets, ready to start with an ice-free bucket in the morning.

If the birds have access to a pond or lake where the centre is not frozen, keep them off until it is completely thawed out. Birds can enter the open water in the centre, then dive and become trapped and drown under the ice.

*This pool, on a natural stream, has been built with concrete shuttering. The floodgate allows for an occasional heavy discharge into the fox-proof pen. The running spring water keeps the birds comfortably warm (above freezing) even when the air and ground temperature dips below -16°C.*

*Special treats in severe weather - cauliflower trimmings and bread.*

# 17 Goose Behaviour

## Introduction

It is fascinating to spend time with your geese to understand their movements, vocalizations and displays, which are still based on the habits of the Greylag goose.

Wild geese may mate for their adult life, though a small number will 'divorce' and re-mate. There must be good reasons for their faithful behaviour, and it is thought that perennial monogamy is an integral part of their reproductive and migratory habits. Species of birds that demonstrate long-term monogamy usually defend their females whilst they are on the nest, will both protect and defend their young, and are always on the lookout for anything threatening. These co-operative strategies work well for the pair, the family group and the flock, and can still be seen in the behaviour of domesticated birds.

## Pairing

Wild geese live in pairs and family groups within the flock, because co-operation is essential for their survival. Well paired geese will stay together for many years, if not for life. This is a successful survival and breeding strategy: they claim the best nesting sites because no time is wasted on lengthy courtship and display each year. Time in the short northern summer is at a premium for these migratory birds: eggs must be laid and hatched, and the goslings sufficiently mature to fly and migrate before the northern daylight hours and the food supply fails. Pairs that choose their nesting site and lay eggs first will produce early, strong goslings, well fledged for the autumn journey south. Late-hatched birds in migratory flocks stand a much lower chance of survival: with insufficient eating and growing time, they may fail to build up their feathers, bodyweight and fat reserves, and may not make the long journey south to the wintering grounds.

Family bonds are longest and strongest in migratory swans and geese, and this must have an evolutionary advantage at the nest. Geese are often forced to nest in quite vulnerable sites, so the attendance of the gander near the goose is a deterrent to predation; two parents are more effective at reducing predation on the eggs and the goslings in their most vulnerable first few weeks.

Attendance by the gander on guard, on the lookout for anything that might threaten the family's safety, also gives the female more time to eat and regain condition: a goose typically loses a third of her bodyweight during incubation. The strongest ganders have females which can graze for the longest periods; single individuals and weaker pairs are more nervous and will move on.

Thus pairing is a good survival and reproductive strategy from the point of view of rearing goslings, and ensuring the health of the female who can also invest in protection of the young. Although geese lay fewer eggs than ducks, goslings usually have a higher survival rate as both parents are there to look after them, whereas in the mallard and many other duck species, the female alone rears the offspring.

Parent geese remain in attendance on their brood until they are fledged. Even after migration to the wintering grounds, the birds will still stay in family groups until, closer to the next breeding season, the adult gander will eventually drive the juvenile males away.

### Pairing Behaviour

Just as geese pair in the wild, domestic geese usually remain paired. In the breeding season especially, ganders will fuss over and guard their goose, putting themselves between their chosen mate and any perceived threat. They will constantly stretch their neck out in front of their mate, and call if they think another gander is more attractive to her.

Well paired birds will always stay close to each other when they graze and rest; they synchronize their movements. However, quite often one in particular will follow the other more closely, and this is generally the gander following the goose in early spring.

A well paired couple of geese will protect their relationship by occasionally driving off other birds in the group. A gander will frequently do this with perceived rival male competitors, because there are often surplus males in wild flocks. After the initial aggressive rush at an offending male, the dominant gander (or even both

*This is my goose! The gander is guarding his goose against the perceived threat of another male.*

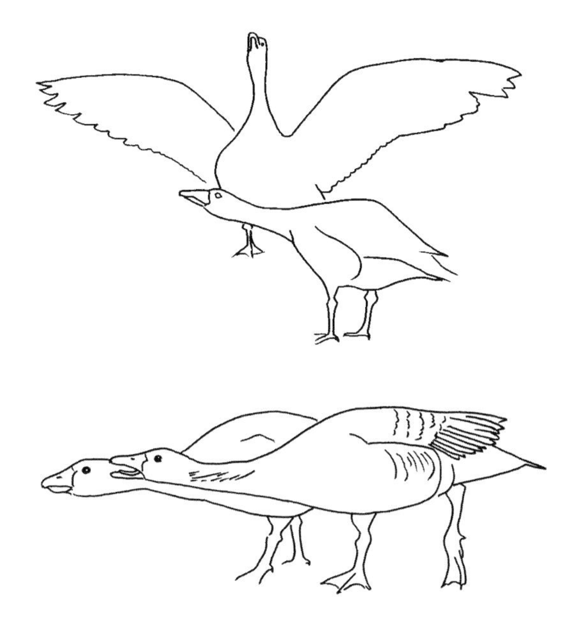

*The triumph display: (a) After a fight, the gander runs back with outstretched wings to the goose. (b)The pair vocalize together, cackling in turn one.*

ganders) returns to his mate to engage in a triumph ceremony. A well bonded pair will call in alternate voices, standing almost side by side, the gander holding his head higher than the goose. Dominant pairs engage in this activity more than low-ranking pairs, and it is a sign of very secure bonding.

A similar activity also takes place in family groups where after a peripheral fracas, goslings and parents (or even a group of goslings if reared without parents) will also perform a triumphal group bond.

## Management Consequences of Pairing

Domesticated geese are best kept in pairs because of their ancestral behaviour. However, young birds may accept a trio or quartet (groups with one gander and two or three females) as long as all the females are introduced together. It is unwise to bring a new female into an established pair because she will be disliked by the resident goose. Never shut up a new goose with an old pair: she may well be picked upon and even killed in a confined situation where she cannot escape.

Even in a well established trio or quartet, the gander will definitely have a favourite goose. As he gets older, other females become neglected, or are even driven away by the dominant female. Larger sets only seem to work with young geese in flocks in the lighter breeds.

Unlike most drakes and ducks, a goose and a gander put together in spring will not necessarily breed. Sometimes there can be instant attraction, but new birds will often take some time to adjust and accept the new partner.

Switching the ganders between breeding pairs of Brecons, for example, can be disastrous. If the birds are able to see their previous partner, they will pace up and down all day, and even out of sight, they will call to each other. Even if they appear to have settled down, and are mating with their new partner, the breeding season can be completely wasted, with no fertile eggs. Back in the flock again in July, the birds go straight back to their original mate. It is definitely best to have well established pairs, so the sooner the breeding pens are together, the better. Geese are best left in their set all the time.

Nevertheless, despite their apparent fidelity, domestic geese cannot be trusted to stay with their partner for mating in the breeding season, especially if there are young males.

## Loss of a Partner

When a duck or a chicken dies, the rest of life goes on. In contrast, geese are curious about death. When a bird dies unexpectedly, well bonded birds (siblings or mates) will stay near the body and talk in low, inquisitive tones, quite different from their greeting or danger calls.

If a bird loses its mate, it is miserable. It may fret and may not eat properly until it has a new companion. Sometimes, after the death of a partner, a new goose may not be immediately accepted, for the gander is still pining for its old mate. Konrad Lorenz considers that 'animals are much less intelligent than you are inclined to think, but in their feelings and emotions they are far less different from us than you assume.' In grieving animals, the general excitability of the central nervous system is reduced. A gander will sit by the body of his old goose in a miserable state; so it is always best to find a new bird as quickly as possible, especially if there is no flock for company.

## Breeding Season Behaviour

### Mating

Washing assumes a new significance in the breeding season, in preparation for mating. The birds definitely get more interested in water, and the ganders proudly preen out their breast feathers by roughly combing them with their bill. Even by late December, just as the days begin to lengthen, the gander may start to 'dip' to the goose on a pool, quickly immersing his head in the water and raising it with a sharply bent neck to attract his mate's attention. At this stage she is not usually interested, as she is not due to lay for several weeks: the gander is behaving in this way to ensure that they stay together as a pair. He will slowly pace about, his neck arched, as he keeps an eye on any competition.

Wild geese generally mate only when they have arrived at the northern breeding grounds, and copulation has been observed in a short period of activity, only two to three weeks before lay. When it is getting nearer the time for the domestic goose to lay, she will also begin to take more interest in the gander's attentions: for example, when he 'dips' on a pool, she will reciprocate by arching her neck and repeatedly dipping her head in the water, synchronizing her actions with the gander. Any gander not giving her enough attention will be roughed up by having his back feathers pulled and being beaten by her wings until he shows interest in mating.

During the act of mating the cooperation of the goose is essential. Unlike other birds, waterfowl have a penis which is extended from the cloaca for copulation. This has probably evolved because of the water environment, to ensure impregnation of the sperm which is directed along the organ, not through it.

After a successful mating the gander falls off with a cry; the female then washes herself vigorously.

*Mating: (a) Ganders arch their neck sharply when they are getting interested in the breeding season; nearer the time of lay, the goose will reciprocate. (b) The goose allows the gander to mate; her cooperation is essential. (c) The gander falls off with a cry that he has succeeded; the female washes herself vigorously.*

### Fighting in the Breeding Season

Ganders may scrap at any time of year, but fighting is much more frequent and serious in the breeding season. If the individuals are evenly matched and the geese well paired, the scrap is usually just a quick flurry of feathers and a rapid separation. If there is great jealousy between the birds, one individual, using his beak, may hang on with great determination to the feathers at the back of his rival's neck. At the same time the opponent will grasp his rival in the same manner, and the birds will beat at each other with their wings. On the wing butt there is the hard, bony projection of the carpal spur, and this is used to flail the opponent whilst the other wing is held back for a better balance.

*An African and a Chinese gander fighting. The birds are exactly the same wild-colour, just different in type. The ganders grasp each other at the base of the neck; the tail is used for balance and the wing is drawn back to beat the opponent – a blow which can numb. Mike Ashton*

*A beaten individual runs off; the victor will return to his goose in a triumph ceremony.*

The ganders will continue to fight, sometimes adjusting their beaks to get a better grip, until one bird gives up, exhausted. The scrap often ends with both ganders out of breath; the heavier bird is more likely to win. The victor may hang on to the back feathers of the loser, but usually he is allowed to escape. Fights can generate a ring of spectators, but other birds will rarely join in, except vocally. As in courtship and mating behaviour, domestic geese mimic exactly the behaviour of the wild Greylag.

## Same-sex Pairs

Once people have bought a 'pair of geese' they tend to assume that it is correct. However, mistakes can be made. Observation of goose behaviour will indicate if anything is amiss, because although humans may have difficulty in deciding the sex, mature geese of normal

*In the absence of females, ganders may start a courting ritual and will rouse each other, but eventually a fight will break out.*

behaviour do not. A goose may scrap with a new goose, but a gander will instantly challenge a new gander, and they will fight if the newcomer does not back off. Even though a gander may not like a goose, he will never seriously attack her. The only problem arises if two young males 'pair up', but they usually give themselves away by their high profile, bossy behaviour with other groups of geese.

Ganders brought up together generally do not fight, and can develop homosexual behaviour. In the absence of females some ganders can be kept in 'pairs' and will live together fairly amicably for most of the year. During the breeding season, however, they will start to go through the courtship ritual, rousing each other by delving the beak into the back feathers, but this usually breaks down at the point where one will fail to submit to the other, and the situation ends in a fight. This behaviour will therefore indicate that you do not have a pair. A female can behave like this towards a gander if she is not getting enough attention, but she then flattens her body to accept him.

When the 'pair' are both females, it may be more difficult to spot. One will appear to 'mate' with the other – the 'mated' goose may even lose feathers where she is grasped on the head – but there will be too many eggs. The laying cycle is about thirty-five hours, so a goose generally lays on alternate days. Observing the behaviour of the geese is important.

## Aggressive Geese

During the breeding season some ganders become much more aggressive. Birds that have lived quite happily together for most of the year can pick on one another, and a dominant individual can make life miserable for both gander and goose in another pair. If they were originally intended to be flock-mated, these dominant birds must be given a different breeding pen otherwise they will upset too many birds, which need a quiet place to mate and lay.

*Characteristic aggressive threat pose, with neck lowered.*

Some ganders become more aggressive in spring with humans, too, but I have never found a bird which could not be persuaded to behave reasonably, using the right techniques. The worst birds are those previously badly treated, or those which are very tame and have no fear. Lorenz also found this to be the case with the Greylag geese that he studied when he approached nests in the breeding season.

Where a gander puts on a display of aggression, ignore it and carry on slowly, as normal. This gives him the opportunity to behave as if he has made a mistake, and give up, or pretend that he has seen you off and return in triumph to his goose. The worst thing you can do is to run away, as this is a clear invitation for him to follow and establish dominance over his terrain. Each time this happens, the habit becomes more ingrained, and it can lead to the bird making contact with beak and wings.

Try not to touch a bird which threatens you. If you catch hold of an aggressive one he may join in with a fight, gripping with the beak and flailing with his wing butts, as if you were another gander. Striking a bird will have a similar effect: it will raise his adrenalin and make him even more eager to fight, whilst you risk injuring him with a random blow. He will also want another fight next time.

If a gander is obviously intending to make contact, try turning to face him and raise your arms: if he is used to being driven, this may have the desired effect. If this doesn't work, try putting your head down and talking: this is a less aggressive pose and the gander may respond by standing and arching his neck, and treating you as an equal. Head-to-head conversations take place with geese which are confident with each other, and you are showing him the same familiarity.

With totally unresponsive birds which insist on going for your legs or face, the only recourse is to make contact: get his neck in one hand and a wing in the other before he is able to strike or bite. Still holding on, turn him round and flatten him on the ground by stooping down and using the inside of your knee if necessary across his back to restrain him. Do not use your weight, but hold him like this for as long as you like: this will establish that you are boss. You are now the dominant 'gander', for a gander which is soundly beaten will lie prone on the ground in a submissive pose until he summons the energy to fly off.

Release him with a final stroke on the back and tweak of the tail, which does not hurt, but represents a final ignominy. This procedure might have to be repeated a couple of times, but it will work: the gander should eventually treat you with respect. Fortunately geese are not like cockerels, some of which never give up.

*The dominant gander is showing who is boss: the male underling will be allowed to get up and run away eventually. This is what should happen when a person holds down a dominant gander: it is important not to hurt him physically.*

This kind of extreme behaviour is unusual and should abate after the breeding season. If it persists, and the gander likes bread, then you can try another ploy. Wear a glove whilst you hold the bread, and allow him to bite it in your hand. With persistence, he will come to prefer the bread to aggression, and he will come to you looking to feed rather than to fight.

## Flock and Family Behaviour

### Imprinting and the Family Bond

The imprinting of the gosling on the mother or surrogate which first looks after it is one of those attributes that make the goose such a sociable and interesting bird. Goslings that are hand-reared, or reared with a tame goose, will readily follow their handler, making them very easy to manage. Such birds also easily adapt to new owners. A group of goslings brought up by a human will imprint partly on that human as well as each other. Treated well, they retain a confidence with people for life.

*Goslings are followers of their keeper, but imprinting must take place in the first few days after hatching.*

Imprinting is a learning process where an attachment develops during a sensitive period in the first few days of life. In that time, goslings learn to follow each other and their mother, and thus secure essential information for their survival. Goslings in a group are much safer than on their own. Their constant peeping, to cement the group, was of course epitomized by Konrad Lorenz in his idiom: 'Here am I, where are you?' Protection by the parents and safety in numbers are crucial to their survival, and this strong bonding instinct carries on over the years as well.

One suggestion for the existence of the long-term bonding of the family group is that the parents lead the youngsters back to the best wintering grounds, and they also take them back through the staging posts to reach the breeding area in the spring. Thus the parents are able to show their offspring the best places for food and security. It is, however, argued that the whole flock could fulfil the same function, but Owen (1980) cites research which shows convincingly that it is rather the strength in numbers of the family group which is the more important factor in the family tie.

The family bond is very important for geese, for the larger the family, the better the territory it can claim. In conflicts for resources, mature birds are more powerful than young birds, and the large family group is more powerful than a smaller family or a pair. Thus larger families dominated the best feeding areas and advantage was conferred: they were heavier, fitter birds.

Wild geese are not mature (for breeding) until their second or third year, and yearling juveniles have not achieved their full body size. Migratory geese actually stay with their parents until the following spring, and yearlings will even rejoin their parents in the second winter. Clearly, it is in the interests of family groups to stay together, and geese will do their best to regroup, even after severe disruption through hunting pressure.

Imprinting of the goslings in the first week of life makes sure that the family keeps together for protection by both parents, but also ensures longer-term better nutrition and well-being, and a fitness to breed as well.

### Flock Behaviour

Wild geese characteristically migrate not only on a seasonal basis but also on a diurnal routine over much of the year (rather like commuting) to feed when visibility is good, and to go back to their safe overnight roost. As flock birds, there is safety in numbers, thus it is important for the individual to stay with the group.

Although domesticated geese have lost the power of sustained free flight, they still remember the wild pre-flight signals, and the flocking instinct. Within a flock in winter, certain birds are always on guard; this is one of the advantages of living in a group. Birds living on their own as singles have less time to eat because of the greater time they have to spend on watch.

Even though domesticated geese do not usually fly, they will still go through the wild-bird pre-flight routine, particularly on brisk, windy days. Signals of reasons to move generally entail a bird looking into the wind with an erect posture, holding the head and tail high, and then making a series of head shakes and shrieks. Any bird may signal this intention by initiating this behaviour, but a flock will not typically follow until there is consensus, and the lead gander also joins in. Preparation for flight will involve others facing up the same way, joining in with the call (which reaches a crescendo in a wild flock) and, at the indication of the lead bird, taking (or running) off.

## Individual Greeting

Birds that know each other, and their human keepers, always use greeting as part of the family bonding process, and it is very important to maintain this with domesticated birds. A group of tame goslings will rush to their keeper with their necks outstretched, and vocalize a series of greetings to establish their bond. To make the birds tame, the goslings also need a response, so it is important to bend down and talk to them head-to-head: the noisier the display, the better. Chinese goslings, in particular, are very responsive and love to be treated in this way.

Birds which have been separated – for example, when only one has been taken to a show – will make a big display to each other when they are reunited, and this can happen if they have become attached to humans, too. We were flattered when we left our first two tame geese for ten days in the care of someone else: when we returned, the two birds spent several minutes in a greeting display of outstretched necks and calls, showing clearly that they recognized us and that they were pleased we were back. This greeting behaviour of the domesticated birds derives from their wild relatives, where displays of intention are critical to their way of life.

FOOTNOTE

For more information on animal behaviour and the modern discipline of ethology, do read the works of Austrian biologist Konrad Lorenz. Ethology is concerned with behavioural processes, and Lorenz chose, for one aspect of his study, a group of Greylag geese. He looked at their social and sexual behaviour and communication, and this led to the production of his wonderful book *The Year of the Greylag Goose* – not to be missed by the goose enthusiast (*see* Bibliography).

# Part VI
# Breeding and Rearing Geese

## Introduction

It seems to be a common complaint that geese are not as good as they used to be. Maybe there is some validity in this view. In the earlier part of the century, geese were more commonly kept as farm birds or as exhibition birds by professional breeders, and only the latter birds were frequently seen at the shows. Now that geese and ducks have become more popular on smallholdings and as pets, which do get to the exhibitions, stock is bred and reared in a far greater variety of ways. The birds reared by the professional breeders with years of experience still make the size they should, but quite often stock in the medium- and heavyweight breeds is too small. Unless good practice in breeding and rearing is followed, then geese do get poorer by the generation.

Geese are much more difficult to breed than chickens and ducks. Most lay only a limited number of eggs between February and June, and are quite choosy about their partner. Young females are often not successful in their first season because their eggs are too small; they need time to grow and mature. Note that wild geese often do not breed until their third year, and that Greylags only lay three to nine eggs per clutch, the number depending on their location and nutrition.

Closely related birds may not produce fertile eggs. Even with a pair of well matched birds, fertility may not be high, whereas in chickens and ducks, nearly every egg can be fertile. Furthermore in pure breeds of geese, fertility can be quite erratic. So don't expect to breed a lot of goslings from a pair (other than from an unrelated pair of lighter breeds, which are easier), especially in their first season.

Pictures of geese out with their goslings look idyllic, but mortality can be very high: Konrad Lorenz cites high losses of Greylag goslings before they are two weeks old by predators and parasites. Equally domestic geese do not always produce a lot of offspring each year, so every gosling is precious, and it is important to protect and care for them very carefully during this crucial stage.

All in all, goose production is a very specialized business. It is skilled, highly seasonal, and you do need a love of the goose. Nevertheless in commercial terms goose has become a popular organic choice for the UK Christmas table market, and the price per kilo is many times greater than the broiler chicken, slaughtered at only forty days of age. Geese for the premium Christmas market can be eight to nine months old, so the costs of production are obviously higher. But this is a luxury food, and in commercial terms, an appropriate conclusion to the cycle of breeding, rearing and finishing geese.

# 18 Eggs and Natural Incubation

## Introduction

As we have seen, most geese lay only a limited number of eggs, and young females are often not successful in their first season because their eggs are too small; also fertility may not be high. Success in hatching is also variable. This applies especially to the UK where insufficient water loss from the egg is often a problem. Close observation of the conditions for incubation, and development during incubation, is needed. Small still-air incubators are not particularly successful at incubating goose eggs of certain breeds, so natural incubation is recommended for small numbers. The microbiology of the nest is crucial for success with early incubation.

However, a small still-air incubator to start eggs to see if they are fertile, and to be used later as a hatcher, is absolutely invaluable. In the 1930s, Reginald Appleyard also recommended that goose eggs were best started in the incubator, but moved on to broody hens which were also safer than the goose for hatching the goslings. A still-air incubator is the best investment to make, to use in conjunction with the goose. It is relatively cheap, and will not become redundant even if a larger automatic incubator is purchased later on.

Commercial goose producers must be successful at using incubators otherwise they would go out of business. So what are the secrets of their success?

Firstly, commercial geese are usually white Embden crosses, sometimes crossed with Chinese. Cross-breeds are easier to hatch than pure breeds, and Embdens were originally praised, in the nineteenth century, for being early layers and the producers of good hatching eggs. When we bred Embdens we also found them relatively easy to hatch in the incubator compared with Brecons and Pilgrims. In general, white geese lay eggs that are easier to hatch in this way than coloured geese, except for Steinbachers and Chinese. These also hatch relatively well in incubators.

Secondly, commercial flocks have genetic diversity: they could be considered the modern equivalent of 'landrace' geese (*see* Chapter 7). These commercial flocks are bred for uniformity of looks, weight and growth potential, and most of the eggs will require the same conditions. Also, commercial geese often lay smaller eggs than exhibition birds, and small eggs are easier to hatch than large ones because their surface area is greater in proportion to their volume and so they dehydrate more easily. This is why there are sometimes better hatches with first season geese than older birds. These goslings do not, however, make the best breeders.

Lastly, the large-scale forced-air incubators as used by many commercial goose producers can be much more tightly regulated for temperature than a small model. The humidity can also be regulated by a dehumidifier or humidifier.

So if you want success with an incubator, get a pair of average, unrelated white geese – but don't breed large numbers of birds if you don't know what to do with them.

## The Breeding Season

Those intending to breed from their geese might be advised to compile a checklist in readiness for the breeding season. In particular they should be sure of the following:

The birds should behave as if they are a pair, and not of the same sex.

* It is important that birds have had several months to settle down together – so you are sure they are well paired – and that they are comfortable in their surroundings. If they are harassed by farm dogs or children the goose cannot be expected to lay well, let alone sit.
* If birds are run as a flock, check that the ganders are not interfering with each other in mating.
* Be sure that the fencing keeps the pairs separated, to avoid cross-breeding.
* Birds should have access to coarse sand, and mixed poultry grit (*see* Chapter 16).
* They need to be fit, not fat. In particular their legs and feet should be sound.
* They should not be too young (yearling geese lay small eggs) or too old.
* It is important that the grazing is clean, with short, high protein spring grass.
* Breeder pellets should be offered as well as wheat

(*see* Chapter 16). Good nutrition is important for both male and female, and essential for high egg production.
* A pool should be provided for heavy breeds to mate.
* Watch the goose for signs that she is starting to lay. Some lay in secret nests under hedges, or dump their egg in the pool. Eggs are laid at the rate of one every other day, and occasionally on consecutive days.
* Given a good diet, the biggest factor in successful goose rearing is the genetic compatibility of a pair – that is, they should not be too closely related (*see* Chapter 3).

## Starting to Lay

Chinese geese amazed their American owners when they arrived in the USA in 1788, choosing to lay some eggs in the autumn. Owners of farmyard geese, too, have reported that their birds have laid in October and November – though these birds may have been brought into lay by artificial night-lighting in the farmyard.

Geese normally lay in spring, and the traditional saying is that a good breeder should lay by St Valentine's Day, when light levels reach ten hours daily (*see* Chapter 21). Commercial birds are crossed with Embdens, which start to lay from mid-February. Romans and Bohemians ('Czechs') are also early layers, but other breeds frequently wait until March and even April before commencing. The older the goose, and the later the cold snaps in the weather, the longer the goose will wait.

## Providing a Nest

If a pair of birds is kept in a small 4 × 4ft (1.2 × 1.2m) shed, there is not much room for a nest. The floor litter must be cleaned frequently to prevent soiling of the eggs if they are laid in the shed overnight. However, if the shed door can be left open inside a fox-proof pen, this set-up is ideal for a nest: the eggs will stay dry in spells of wet weather, and the open door can be screened off from crows and magpies by cypress branches; these will also protect the goose from the sun in a heatwave.

Nesting material must also be provided. Chopped flax, barley or wheat straw is good for this purpose, but not hay, because it goes mouldy. On a hard shed floor, the nest litter must be very deep. The goose will try to scratch out a hole in the floor, as she would do naturally on earth. Eggs break on a hard surface, so a base of old carpet covered by plenty of peat or shavings makes a good start. The edges of the nest need to be supported by heavy pieces of wood as retaining walls, or a large container such as plastic garden trug employed.

*A nest lined with down in a safe place. The goose usually covers her eggs when she leaves to feed.*

A large piece of turf, moulded and beaten into place, also makes a good base. The central hollow can then have chopped litter added, and when the goose goes broody, she will line the nest with her own down. Make sure there are no holes for eggs to roll into and get lodged – or goslings to get stuck inside if they are to be hatched in the nest. The nest must be a smooth, concave shape with no obstacles.

If the birds live out in a fox-proof pen and there is no

*A wigwam of cypress branches is good for keeping the nest dry. Help the goose to make a scrape in the ground and provide enough nesting materials so that she can hide her eggs.*

shed for them to use, encourage them to use the best nesting sites. You cannot move a goose from her chosen place, so it is best to develop a nest site which is dry and well concealed. Help the goose to make a scrape in the ground in a fox-proof pen: loosen the earth, and she will mould it with her feet as she turns around, and provide nesting materials so that she can hide her eggs.

The eggs may be vulnerable to predation by rats and birds, and it is best to collect them from the nest on a regular basis, rather than let the goose sit, unless it is well concealed. Any goslings hatched outdoors are generally instantly eaten by predators - rats, stoats, birds, foxes - if the eggs have not been taken from the nest already.

## Eggs and Egg Production

Birds generally have one ovary which contains several eggs in different stages of development. In spring the ova prepare, in sequence, for ovulation into the oviduct by developing a yolk over a period of several days.

The oviduct is a tube which extends from the ovary to the cloaca (*see* diagram). It is open-ended, and a ripe yolk (ovum) must be 'caught' in the mouth of the infundibulum so that it does not pass into the peritoneal cavity of the abdomen. A bird in lay must be treated quietly so that her eggs are laid normally, in the oviduct.

Fertilization occurs when the ovum is released into the oviduct, which is only a short window in time, around fifteen minutes: to achieve this, females must store semen in the oviduct in sperm storage tubules - blind-ending tubes in the oviduct wall. Sperm make their way from these storage areas up to the yolk, against the flow of the egg, and then penetrate the germinal disc. A sperm can pierce the vitelline membrane enclosing the yolk, but is unable to penetrate albumen once it starts to form. The storage period of viable sperm varies with the species: in ducks it is up to eighteen days, but it seems to be considerably less in geese.

As the egg passes down the oviduct, a coating of albumen is added in the magnum, and the spiral chalazae anchor the yolk into its central place. The spiral develops as the egg rotates in its passage down the oviduct. The membrane is added next, in the isthmus, followed by the shell in the shell gland. The mucus of the lower oviduct also coats the egg in an anti-bacterial layer. Once laid, the mucus layer dries very rapidly, and if preserved intact, is a very good defence against bacteria. The whole process takes around thirty-four hours in the goose.

During this time - when the egg is formed and laid - a fertilized egg (the zygote) undergoes cell division, and the embryo is then ready to continue development when incubation commences. The fertile egg is a living structure and must be treated carefully.

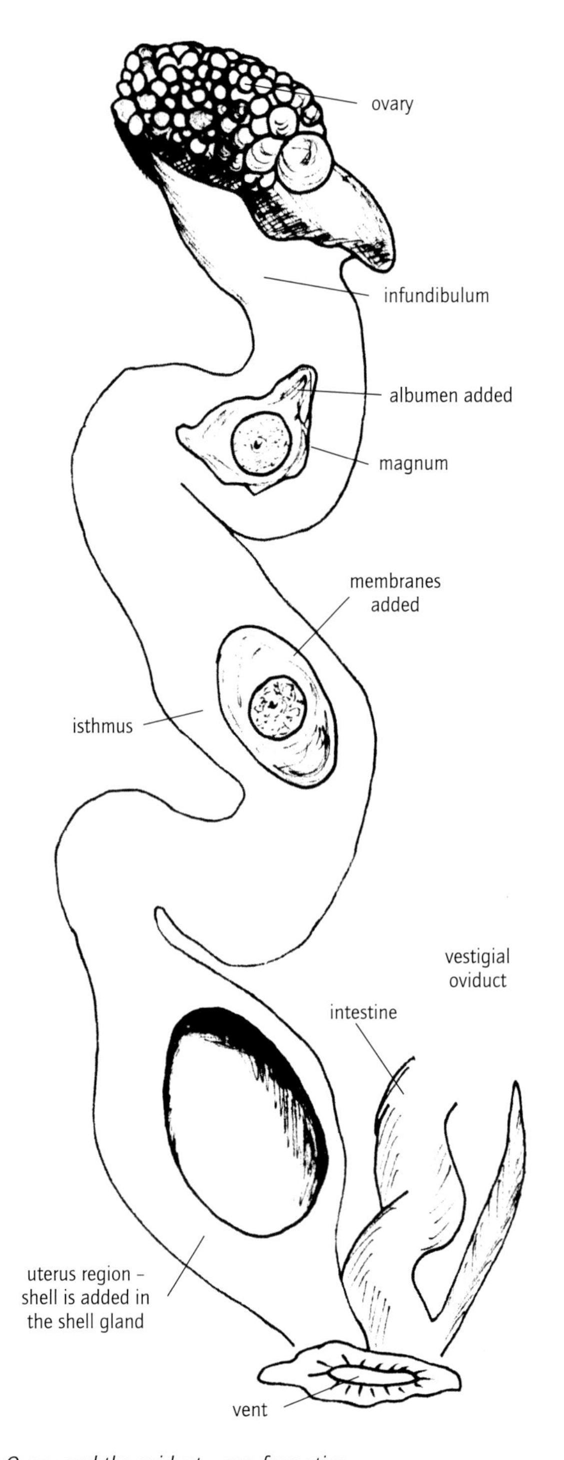


*Ovary and the oviduct - egg formation.*

## STRUCTURE OF THE EGG

The egg is covered externally by a porous shell formed from the external cuticle and the chalky testa. This chalky layer gives the egg its strength and provides the calcium that will be partially absorbed by the embryo. The cuticle blocks openings in the pores of the $CaCo_3$, but still permits gas exchange.

Inside the shell is a membrane: it has two layers, which separate at the broader end of the egg to form the air sac. The outer membrane is attached to the testa and the inner membrane to the albumen. Gases are exchanged between the albumen and the blood supply of the embryo and the exterior through the porous shell structure.

The albumen is held within the membrane, and within this are two twisted cords, the chalazae, which are attached to both sides of the yolk sac. Their function is to suspend the yolk in the albumen and allow it to rotate freely. These cords are coiled so that when the yolk rotates in the thin albumen, one coil is wound up as the other unwinds. Turning the egg in the same direction all the time will therefore destroy the egg's structure. Most of the albumen is dense in structure and extends to the ends of the egg where protein fibres are attached to the shell. The inner and outer part is fluid albumen, the inner fluid layer allowing the yolk to rotate on turning. The albumen is a store of protein and water and is used up during incubation.

The yolk is rich in fats and proteins and is surrounded by a thin membrane. The yolk itself consists of alternating dark and light layers: these are laid down by night and day respectively. The centre of the yolk contains a ball of protein-rich yolk, a narrow column of which extends to the surface where it broadens into a cone. The blastodisc (germinal disc) sits on top of this. The majority of the yolk is not used during incubation, but is drawn into the abdominal cavity of the gosling just before hatching.

Whatever position the egg occupies, the blastodisc occupies the upper position. This is a small white spot (about 4mm across) from which the gosling will grow. The part of the yolk next to the germinal disc is less dense than the rest of the yolk so it tends to float upwards, rotating the yolk so that the germinal disc is always on top. In the early stages of development the embryo can only use the nutrients in contact with it, so turning the egg gives it a new source of food and oxygen from the inner liquid albumen.

The yolk and the germinal disc originate in the ovary. The albumen, shell membrane and shell are formed by the oviduct.

(Adapted from Charnock Bradley, 1950 and Brooke & Birkhead, 1991)

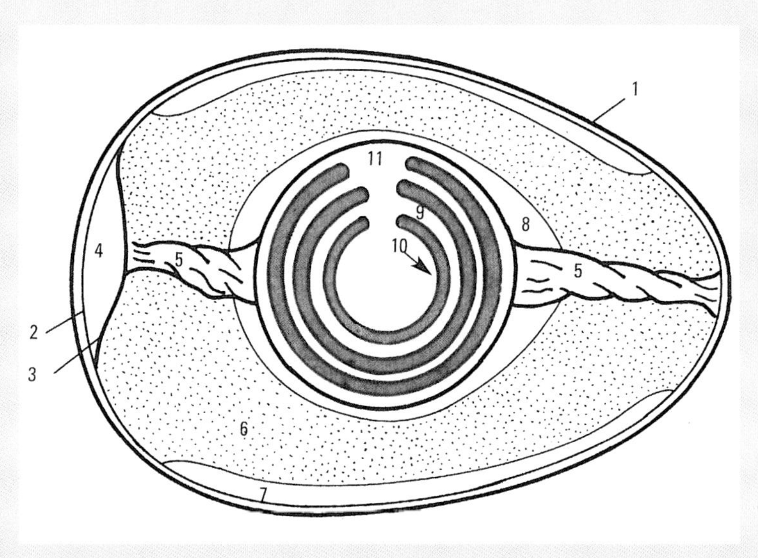


**1 shell**
**2 shell membrane**
**3 egg membrane**
**4 air sac/cell**
**5 chalazae**
**6 dense albumen**
**7 outer liquid albumen**
**8 inner liquid albumen**
**9 white yolk**
**10 yellow yolk**
**11 blastodisc/germinal disc**

## Caring for the Eggs

### *Collecting, Labelling and Cleaning*

Most domesticated geese lay more eggs than they can brood (*see* Chapter 3). Eggs left in the nest can get very dirty, so collect them regularly. They can be laid at any hour, and are readily lost to predation. Label each egg in pencil with the date and parent to discover which birds are having a successful season and which are infertile. If there are several to collect, carry a bucket with some wheat in the bottom, or shavings, so that the eggs do not roll around. Very dirty eggs should be discarded.

If eggs need washing, it is important to use tepid water; cold water causes the egg to contract and draw bacteria in through the shell. Any eggs laid in water are useless. Use running, tepid tap water to wash particles and bacteria away. Take care not to keep turning the egg in the same direction whilst cleaning, as this will affect the internal structure of the egg. Also, handle the egg gently otherwise the embryo may be damaged. If eggs are immersed, use a drop of bleach such as Milton, or a proprietary egg sanitizer. When well dried, record the weight of each egg on the eggshell. The fertiles can then be monitored over the incubation period for their loss in weight.

If the eggs are not unduly dirty, and the mud or droppings are dry and will brush off easily, they are best left unwashed, especially for natural hatching. The mucus from the oviduct which helps the goose to pass the egg also forms a protective, smooth coating which helps to exclude infection. Fewer goslings will become ill if this cuticle is maintained as a barrier against harmful moulds and bacteria.

### *Storage*

Temperature, humidity, position and turning are all important factors when storing eggs.

**Temperature:** Whilst the goose continues to lay, the eggs should be stored at 12–14°C. The egg is dormant at this temperature; development commences above 21°C.

**Humidity:** Water is lost through the pores of the shell in a dry atmosphere, especially if the air is moving. A relative humidity (RH) of seventy-five is recommended for storage.

**Position:** Storing eggs on their side requires a lot of space, so the eggs can be placed vertically, propped up in a corner against a wall, with the air sac uppermost. The blunt end uppermost is often recommended, but it is also argued that the pointed end up is better as there is more thick albumen between the yolk and the inner membrane (to avoid the germinal disc making contact). Note that, in one species that uses a vegetation mound to incubate the eggs, only those lodged in the vertical position hatch. The eggs are not turned, and the air-sac-down eggs never hatch. Storing eggs vertically is all right for a week; after that, store them on their side and turn them daily.

*Storage: these eggs were labelled and dated when they were collected. Storage in polythene bags is recommended for dry climates. Preserving the $CO_2$-rich environment of the egg may also be beneficial.*

**Turning:** Eggs stored on their side should be turned once a day through 180 degrees, in the opposite direction from the time before. The albumen acts as a shock absorber and suspensory mechanism for the yolk, but left in the same position for several days, the lighter yolk moves upwards and sticks to the shell's membrane. This will adversely affect the germinal disc.

## Selecting Eggs for Incubation

For the incubator, eggs should ideally be two to ten days old; an egg that has only just been laid should be cooled for at least a day.

For natural hatching the eggs must be no more than twenty days old. Geese do not lay every day because the cycle is about thirty-four hours; occasionally there will be an egg on consecutive days.

There are often more eggs than can be sat by a broody goose, and the first ones laid may be too 'old' by the time she goes broody. Early eggs can therefore be eaten, or they should be placed under broody hens.

Not all eggs are suitable for incubation. A goose may lay thin-shelled eggs, eggs with a 'chalky' feel, which

*Selecting the eggs: the irregular shell on the right will be too porous. The two pink eggs (left) indicate a bird having problems in the oviduct; these are likely to be infertile.*

seem very porous, or eggs with an irregular exterior. Those with a poor quality shell may be fertile but they will dehydrate too quickly and die. There can even be a succession of eggs with no shell at all, though the bird may lay normally the next season. Ensure that plenty of mixed poultry grit and calcified seaweed is available (*see* Chapter 16).

Reject small eggs - such as from yearling birds - as they may produce small goslings. Large eggs are frequently double-yolked; this can be checked by candling, when the yolks will appear as two shadows. Any cracked eggs will also be revealed; these should not be set as they will go rotten on incubation.

## Natural Incubation

Many people assume that every goose will sit. Some breeds, such as the Toulouse, are unlikely to go broody, whilst others, such as Embdens and Africans, are happy to sit, but can smash almost every egg because of their weight. The best sitters are the medium- and lightweights (*see* Chapter 3). Chinese are less likely to go broody until late May or June.

Allowed to follow her instincts, a goose may incubate her first clutch. This, of course, limits her potential output to the freshest nine eggs in that clutch, and after sitting for thirty days, this generally means that no more eggs will be produced that year. Sometimes, if no goslings hatch, the goose might lay again, but in no circumstances should she be allowed to sit more than once a year.

If all the first-clutch eggs are taken from the goose and incubated under broody hens, she will be discouraged from going broody. A second clutch of about ten eggs usually follows in late April or May, after a break of about three weeks. These eggs are often more hatchable in the incubator, and frequently produce a higher proportion of females. They may, of course, be incubated under the goose, but more attention must be paid to excessive water loss in the nest at this time of year (*see* later section 'Checking the Eggs').

### Encouraging the Goose to Sit

If the goose is to sit, she must have a nest where she feels safe. The nest litter must be clean and dry. To encourage her to go broody, keep the first two, dated eggs in the nest. Remove, date and store the eggs that follow, to keep them clean. Return seven to nine of the freshest eggs when she is clearly going broody. It is best not to have the nest too crowded as it will result in breakages and some eggs being too cool. The first

*Broody geese! More than one on a nest will cause problems, but sometimes it is difficult to persuade the geese otherwise. Valuable eggs are taken and incubated under lighter-weight birds.*
*Hans Ringnalda/Sigrid van Dort*

CONDITIONS IN THE NEST

The average body temperature of incubating waterfowl is 40.7°, and the breast brood patch 39.5°. In the nest, egg temperature varies with the location: air sac temperature was recorded at 35.6°, while nest air averaged 31.7°. The bottom of the nest averaged 25.2°. The temperature of the actual eggs (in a mallard nest) varied by as much as 5.6° between a central and a peripheral position. Mean egg figures for a Canada goose varied from 36.6°–39.7° from days fifteen to twenty-five during the incubation period. Broodies regulate the temperature by adjusting their attentiveness and the 'tightness' of their sit. (These figures are from Afton and Paulus, 1992.)

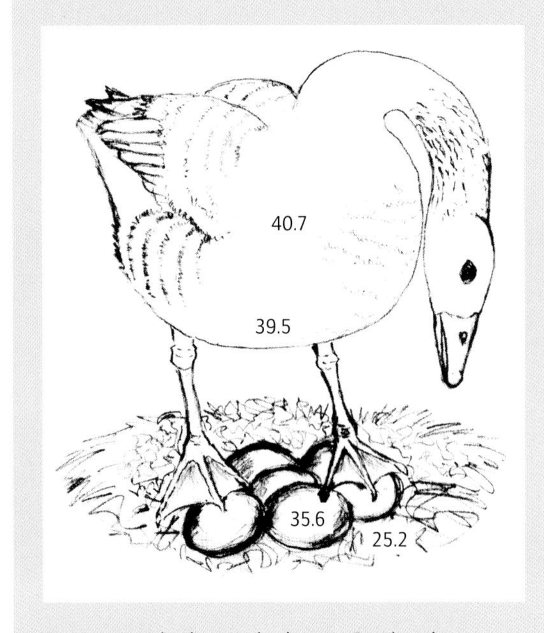


*Temperature in the nest in degrees Centigrade.*

two eggs will get too old and dirty, and can eventually be discarded.

Incubation behaviour, which is dictated by the action of the hormone prolactin, will generally not start until the goose has a satisfactory nest and a clutch of eggs. Prolactin increases steadily with egg laying, but increases dramatically with the onset of incubation. Visual contact with eggs, touching them, and contact with the goslings later on, will all maintain prolactin levels. If all the eggs are removed as she lays them, then broodiness may not develop.

If the goose is ready to sit, she will line her nest with down and then she can be given a couple more of the oldest eggs to check if she is serious about sitting. Frequently a goose will not settle well. If the weather turns extremely cold she may decide to have a break from laying rather than go broody. Don't worry if she has messed about with eggs and sat on them at night and warmed them, and then gone off and left them to go cold the next day, because they will still be all right. In the wild, the eggs would naturally be protected by the goose at night as she comes near to sitting. Warming the eggs slightly on a daily cycle improves their viability, especially in eggs over ten days old.

If there are two geese in the same shed they may disturb each other. If they are close together they will steal each other's eggs, rolling them from nest to nest. Sometimes the two geese will want the same nest, leaving the other clutch of eggs to go cold in the middle of incubation. It is best to keep just a pair of geese, or if you have a trio, to have the sitting females in different areas, and to ensure that they go back on to the right nest. It is therefore important to observe them closely.

## Caring for the Broody Goose

More females are lost in spring through lack of care than at any other time. It is essential to make a note of the date when the goose started sitting, both for the sake of her health and for that of the goslings. The incubation period is twenty-eight to thirty-four days, and it is important to note both its start and end point. Close supervision will be needed around hatching.

When the goose is definitely broody both she and the gander should be wormed. The advantage of worming the goose is that she does not lose so much condition whilst sitting, and one cause of disease is eliminated if she becomes ill. Also, this reduces the incidence of worms when she and the gander lead the goslings out.

The goose must be fed and watered at least once a day. She will carefully cover her eggs with down, so they will remain warm and be camouflaged. Wheat in a bucket of water is suitable. She should be encouraged to swim if the weather is hot and dry; this will give the eggs the correct amount of moisture, and allow her to remove any parasites more easily.

Birds which are sitting on eggs must take time off the nest to feed, drink, preen and defecate. Larger species with large eggs, such as geese, tend to leave the nest just once a day for quite a prolonged period of time. These recesses can be for fifteen minutes, but can be up to an hour, especially in spells of hot weather.

With tame birds, the goose will allow you look at the eggs and lift her off the nest, and a tame gander will not attack. In these circumstances, a goose can be fed twice a day if she is losing condition: check by lifting her off the nest. With fierce birds, this is much more problematical, and it is best to drive the gander away to a secure place whilst the goose is encouraged off the nest, otherwise the eggs could risk being smashed.

Geese which are accustomed to sitting will probably look after themselves. Young birds need more attention because they are inexperienced and can get run down by sitting too tightly. In addition, broody birds can become infested with external parasites. Geese generally suffer less than hens in this respect, but they will need checking. Follow the procedure for broody hens (*see* later section 'Broody Hens and Muscovy Ducks') if in doubt.

Note that wild geese also become very emaciated during incubation, and then must feed intensively in order to regain weight and to moult; they lose more than one third of their bodyweight, and have been known to die on the nest. Therefore look after domestic females very well, or they will be lost. Wild males also lose 20 per cent of their weight because they are having to be extra alert or are defending their territory: they have reduced feeding time and grazing space whilst on guard.

## Checking the Eggs

Eggs in the nest are vulnerable to all sorts of risk. Waterfowl turn their eggs about once every hour; they are generally just stirred around, but the beak is also placed under an egg to hook it end over end, a procedure that can cause eggs to crash together and fracture. Vermin can also cause losses; rats in particular can take every egg from under a goose, so a rat-proof shed is essential. Magpies are adept at swooping in when a goose is off the nest, so the eggs must be well covered.

It is pointless for the bird to sit and lose condition if all the eggs are infertile. It is best to start with clean eggs, because it is virtually impossible to candle a dirty shell (*see* next paragraph and Chapter 19). As incubation proceeds, some of the embryos may fail, a situation which is easy to identify with clean eggs in an incubator, but far more difficult in the nest.

Water is lost during incubation, and the rate of loss affects hatchability. The most accurate way of monitoring this is to weigh the egg, but a quicker way is visually, by candling. The rate of change of the air sac is greatest from day eighteen, and there can be a sudden change from days twenty-five to twenty-eight (*see* the egg on the right in the illustration). Evaporation and the size of the air sac can be controlled in an incubator by varying the humidity, and halting weight loss if necessary.

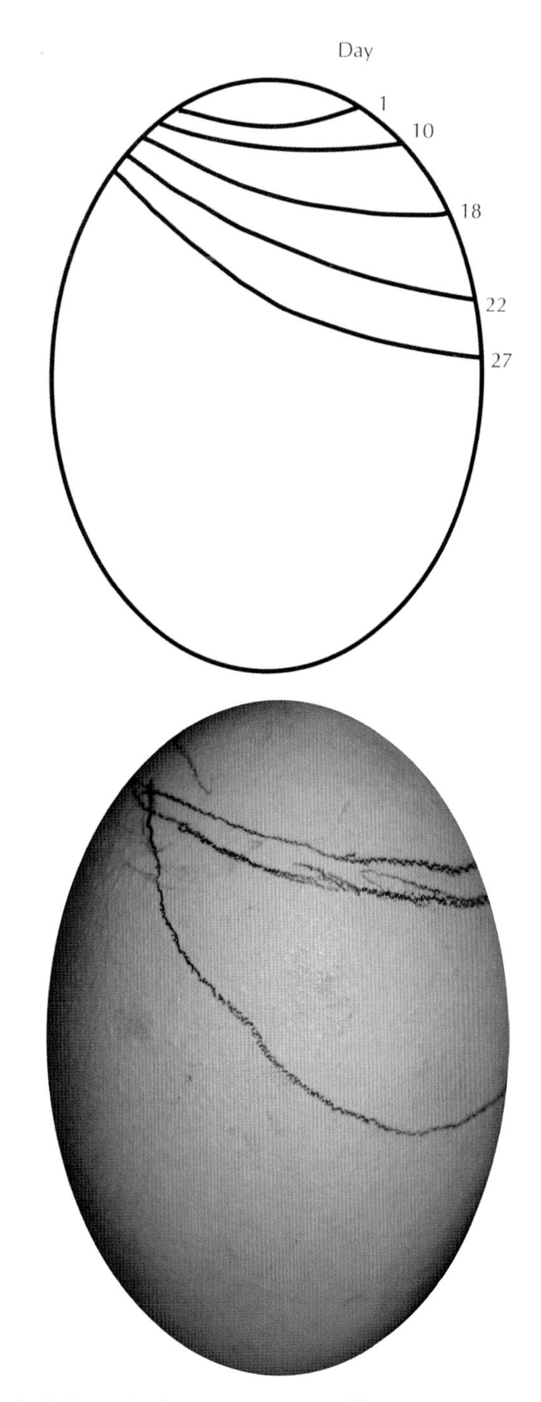


*Ideal air sac development seen on candling.*

Eggs which have started and then failed will go rotten, and these need to be identified and discarded. If, on candling, an egg's air-sac seems non-existent mid-way through incubation, the egg is bad. If you cannot tell by candling, you can probably tell by smelling it – though bear in mind that if an egg has broken in the nest it can transfer quite a smell to the surface of the others.

If by this stage the surfaces of the eggs have become extremely dirty, this does not matter. The natural preen-gland oil from the goose's feathers cleans and polishes the eggs. It secretes a natural antibiotic, as well as affecting the permeability of the shell.

If a dirty egg could be rotten, then shake it. If it is seriously rotten the contents will slop around inside, and it is essential to dispose of it before it explodes in the nest. If the egg is oozing a liquid it should have been thrown out earlier.

Mark the air sac's outline in pencil to monitor progress. In late May and June eggs can dehydrate far too rapidly under a goose and are better relocated to a damper incubator. By contrast, earlier eggs are much more difficult to dehydrate, and incubate far more successfully in the nest.

### Hatching in the Nest

As day twenty-eight approaches, the eggshells will have become thinner as the embryos resorb some of the calcium for skeletal development. The eggs will also feel hotter. The goose will stand up and over her eggs in order to cool them; an experienced goose often turfs cooler, dead eggs out of the nest.

Healthy goslings use their muscle to chip and push at the inner membrane with their egg tooth, and burst into the air sac. It is essential that this sac is well developed, to about one-third of the volume of the egg, so that the gosling can fill its lungs. At this stage, when the goslings start to breathe, they squeak. They next work at the shell until they pip a hole in it, at any time between days twenty-eight and thirty-one. The goslings then rest for at least a day, building up the oxygen supply in their blood, before making the final effort to turn round in the shell, in a can-opener action, and prise off the lid. The goose will assist with this hatching process, nibbling off bits of shell and often devouring quite large portions of the discarded egg. If any eggs are broken during the progress of incubation, the remains can go the same way too, and may explain some losses.

Some females are adept at hatching their young, but others are hopeless. To avoid losses, the eggs can be transferred to the incubator on day twenty-seven or twenty-eight. If the goose has reared goslings before, she will be happy to have these back (as long as she has something to sit on in the meantime); however, if she has never hatched a gosling before, then don't expect her to know what it is all about.

A clutch of eggs can hatch over a two- to three-day period, and up to four days has even been noted in wild nests (Owen 1980). Hatching time will vary according to the age of the egg when set (older eggs take longer) and water loss (dry eggs can hatch earlier). However, hatches often synchronize. The goslings communicate by tapping and squeaking up to three days before shell pip, and the goose communicates with them, too. This activity plays an important part in both the synchrony of the hatch, and on imprinting, and this behaviour must be a huge evolutionary advantage in precocial species in the wild, because it means that all the viable eggs will be protected and hatched by the goose, instead of being deserted and suffering predation.

Communication between the goose and goslings also keeps them together on the nest for one to two days after the start of hatching. The trilling of goslings needing warmth and brooding has a soporific effect; whilst they settle down, the note of their voice encourages the goose to brood them and keep them together. However, the strongest goslings will start to explore, and then return to be brooded, and these forays will become more persistent; also the goose must leave the nest to feed. Late-hatching eggs will be abandoned as the strongest goslings become increasingly active.

During the brooding period the vocalization between goose and goslings bonds the family together, so the young will follow the mother goose using both vocal and visual cues. In the wild, this nidifugous (nest exodus) behaviour means that weak goslings and late hatchers are left behind and taken by predators. Note that most eggs of wild geese do hatch, and that very few are infertile, addled or 'dead in shell' (Owen 1980).

The fluff of goslings hatched in the nest is always in good condition. During the brooding of her young the goose stimulates her preen gland, and oils her feathers. The oil is then transferred to the goslings as they all constantly brush together in the nest, with the result that their fluff is structurally well conditioned and more waterproof than fluff on goslings that have come direct from a hatcher.

Be aware that females left to sit for more than thirty-two days on dud eggs often find it very difficult to revive their appetite with no young to stimulate them, and sometimes die. If the nest is removed the goose will do her utmost to reconstruct another on the same spot. The best strategy is to move the goose to another paddock, or to give her some goslings from another bird if she has reared goslings before.

### The Key to Success

The key to success in natural hatching is good observation. Since this is a natural process, many people assume that the birds will manage themselves. However, these are not wild birds: they have not chosen their patch, their human keeper has; nor have they been trained by their parents in the skills of survival. Also, they may be kept more intensively than wild birds so the density of parasites may be higher. One similarity is that, in the wild, many eggs and goslings are lost to predators, and those in a domestic environment will be too unless their human keepers are vigilant.

Commercial gosling producers are not able to justify such a labour-intensive method of production, but hatching pure breeds is much more difficult than hatching a carefully chosen commercial cross. It is most satisfying to manage stock successfully without too much equipment, and many people keep pure breeds out of interest rather than commercial gain.

## Broody Hens and Muscovy Ducks

Broody hens and Muscovy ducks are more successful broodies than most geese. They are lighter in weight, and can be persuaded to sit earlier, often shortly after the geese have begun to lay. However, their nest must be adapted to the local climate and weather. Thus damp earth nests are not good in March: an airy wicker basket with chopped straw is much more effective. Conversely, in a dry May the nest might need a daily spray of water and a base of damp turf, and dry eggs often need to be moved back into a very humid incubator if they are to survive.

Any type of bird can be used for sitting as long as she has the broody trait. Smaller birds will cover only two to three goose eggs, but bigger hens or a Muscovy can cover five to seven. They are best started off on their own eggs; then, when they are thoroughly broody, goose eggs can be substituted, ideally warm, fertile eggs that have been started in the incubator. Eggs that have spent even fourteen days under a broody have a better hatch rate than those which have been wholly artificially incubated, especially early in the year.

Although hens sit on their own eggs for only twenty-one days, they manage for longer if they are well looked after. The health of the broody hen can be assessed by her weight, the colour of her comb and her behaviour on feeding. The Muscovy will sit naturally for thirty-five days and will also cover a lot of eggs. Birds become very tame when lifted off the nest once or twice a day for feeding and for randomly hand-turning the goose eggs in case they are having difficulty.

Wheat is recommended as a broody ration; the droppings are firmer and the nest is less messy if fouled. Birds enjoy wheat which has been soaked in water, so leave some in a bucket overnight ready for the morning feed. This also helps ensure that broodies do not get dehydrated.

Because she is unable to preen effectively, a broody can become infested with mites so she will need the help of an insecticide. Also, change the nest litter every couple of weeks. Before fresh litter is used, dust or spray the nesting container and the bird, particularly at the vent where parasites accumulate for moisture. Keep the broody clean to avoid infestation of the hatchlings, which must be kept clear of parasites. It is essential to avoid poisoning the goslings with insecticide dust near

*A Muscovy duck covers five to seven goose eggs, and can quite happily sit for thirty-five days.*

hatching, so the use of ivermectin on the sitting hen is now recommended by vets (as Roberts 2000, *see* Bibliography; *see also* Chapter 22).

Hens can hatch and rear the goslings, but because their feeding habits are so different, it is preferable to rear the goslings as a group on their own. Hens can scratch showers of dust over the goslings and persuade them to eat worms; also if no grass is made available, the goslings may even 'graze' the hen and begin to strip her feathers.

## Microbiology and the Nest

It is generally accepted that eggs from pure breeds are best started under broodies because natural conditions are better than incubators; the temperature should be steadier, and of course there is no overheating. However, the improved natural hatching rate with goose eggs is largely because birds provide the right conditions for correct water loss, despite research indicating that there are enormous variations in humidity within nests. The average relative humidity (RH) is around sixty-two, which would be far too high in an incubator in the UK.

Eggs set under broodies in March and April usually lose the correct amount of water, whilst those in the incubator often fail to hatch: they remain too wet, despite the RH falling to twenty to twenty-five. However, the loss of water in eggs in the nest can be far too great from May onwards, when these eggs might have to be rescued and placed in a humid incubator. This is probably because of the huge differences in shell quality between the beginning and end of the season. This phenomenon, of differing rates of dehydration over the season, has been regularly observed over many years.

The hatchability of early eggs in the nest is best explained by the microbiology of the nest. In general, nests are best kept as clean as possible. However, they are naturally very humid, messy places, and harbour fungi and bacteria, some of which are transferred from the bird's feathers and skin. The cuticle of the shell will help prevent decomposition of the egg by these organisms. Deeming (2002) quotes research which has found that *Bacillus licheniformes*, commonly found on feathers, thrives on the cuticle and also excludes other bacterial species: it therefore has a beneficial, hygienic effect. Not only that, whilst the bacteria degrade the cuticle of the egg, the shell becomes more porous to water vapour. Even if transferred to an incubator, this quality remains after a period in the nest. Such eggs, now in the incubator, will require more moisture.

This phenomenon also explains goose eggs losing a huge amount of water in the nest in May: not only do the eggs have a thinner shell at this later time, but the increased bacterial activity in the heat also results in rapid water loss.

Add to this the frequent airing of the nest by the bird raising itself to have a break after a sitting spell, shuffling the eggs around and cooling them as well, and one can see that the messy conditions of the natural nest are quite difficult to mimic in the incubator. The nest is a little ecosystem in its own right, and, of course, superbly adapted for its purpose.

# 19 Incubators

## Introduction

Modern incubators have electronically controlled thermostats which keep a steady temperature from the heating element. Some also have humidifiers to maintain a specific relative humidity (RH). However, reducing the RH, rather than raising it, is more often the problem for early goose eggs with dense shells: controlling water loss is *the* key variable in hatching goose eggs, so eggs with differing properties are unlikely to hatch well in the same conditions in one incubator.

## Types of Incubator

Electric incubators have the disadvantage of being out of action in a power cut, which is why a few people still prefer the older paraffin types. However, these must now be obtained second-hand, and they are much more difficult to operate. There are two main types of electric incubator: the still-air and the forced-air (also called 'fan-assisted').

### The Forced-air Incubator

In the forced-air incubator, a fan circulates the air around the whole cabinet so there can be several layers of eggs. The even flow of warm air ensures that the incubator temperature is the same throughout. These incubators are generally more expensive than the still-air models because they are usually larger, need a fan, incorporate automatic turning, and often have humidity control.

In a forced-air cabinet, the automatic turn is generally through 45 degrees, to the left and to the right, making a total of 90 degrees. This is not ideal for goose eggs, where a hand-turn of up to 180 degrees is better. If an automatic turner is purchased, check that the angle of turn is more than 90 degrees, and also ensure that the design will cope with the weight of goose eggs.

Eggs due to pip can be moved down to the floor of the cabinet for hatching. However, this is not recommended because 'hatching fluff' will find its way into every part of the incubator: it is better to have a still-air hatcher, separate from the clean cabinet incubator. The advantages of the forced-air incubator are its steady temperature and ability to evaporate more moisture (because of the moving air), if required, for weight reduction.

### Still-air Incubators

Still-air incubators have only one 'layer' of eggs. An element in the insulated box heats the air, which convects naturally. The temperature is adjusted so that it is correct for the layer of eggs at their top surface, which is where the tiny embryo lies. Some incubators have automatic turning gear, to roll the eggs left, then right. A small still-air incubator is ideal as a hatcher, especially if it is easily cleaned. Hatching eggs make a mess, so it is best to keep them separate from incubating eggs.

*Brinsea still-air incubators are well insulated, durable and easy to clean. The Hatchmaker is the perfect design for a hatcher – the base can be immersed in a bath, and there are no hidden places for fluff and bacteria to accumulate.*

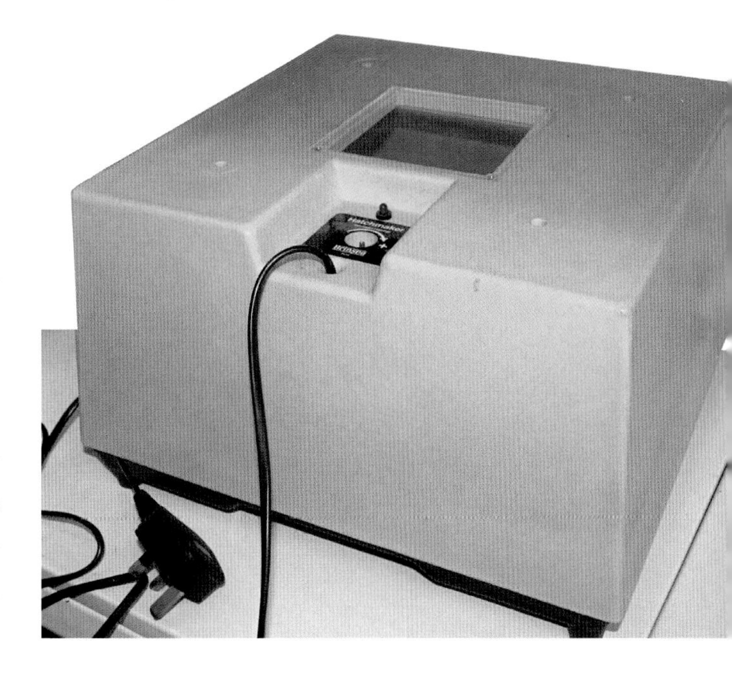

Unless the one incubator goes right through the whole process of setting eggs, hatching and cleaning before another batch of eggs is started, then two incubators are essential, for several reasons. Firstly, if cold eggs are added to a still-air incubator the temperature drops, and it will take hours to stabilize again, which will adversely affect a batch of eggs already started. With two incubators, one can be used to warm up and start a new batch without affecting eggs that are part-way through incubation.

Secondly, eggs run at different temperatures. Eggs near hatching generate their own heat and are warmer to the touch than newly incubated eggs; in a small incubator some of these feel really hot, while others are not warm enough. These hot eggs, which are 1–2° warmer than newly set eggs, are best transferred to the incubator to be used as a separate hatcher, because the air temperature can then be regulated down to 37°C.

Two incubators also give the opportunity to run a very dry and a very wet incubator for specific eggs, should the situation demand it – for example, a humid incubator at hatching.

Thus two still-air incubators, which are simple to operate and have little that can go wrong with them, are very useful. If later you progress to a forced-air cabinet model, the original incubators are invaluable as hatchers.

### CHECKLIST FOR SUCCESSFUL INCUBATION

* Eggs and incubator must be clean: bacteria and moulds can make eggs go rotten, kill the embryo, and cause infections in newly hatched goslings.
* The egg must be turned frequently to bring the embryo into contact with fresh nutrients in the albumen.
* The right temperature is crucial because it allows the correct speed of growth. High temperatures kill eggs.
* The egg must have a supply of fresh air to provide oxygen and to remove waste carbon dioxide.
* The right humidity is also crucial: the egg loses water through pores in the shell, and the humidity of the air around it must be controlled to ensure that the right amount of water is lost over the incubation period. Candling and weight loss calculations monitor this.

*A still-air incubator designed for 150 hen eggs, and holding a mixture of goose and duck eggs. Many goose eggs may prove to be infertile, so it is useful to set quite a large batch at a time and then transfer the fertiles to smaller incubators or broodies.*

## Managing the Incubator

It is important to find a suitable place in the house to run the incubator so that it performs reliably and to maximum efficiency. The following points ensure that the small incubator does as good a job as possible:

* Place the incubator on a level surface.
* Ensure that it holds a steady temperature by locating it in a well insulated room at 18–21°C. A north-facing room is best, to avoid the incubator overheating as a result of direct sunlight.
* The room must be clean and adequately ventilated. A hot room will reduce air circulation by impeding convection in the incubator, and carbon dioxide will build up.
* A small incubator will maintain a steadier temperature if it is partially covered by a blanket, which does not block ventilation. There are holes for incoming air in the base, and outgoing carbon dioxide at the back or top.
* Humidity must be low, in order to control the relative humidity (RH). Choose a dry part of the house: it is easy to introduce water, and much harder to take it away from the environment with a dehumidifier (a common problem in the UK).
* Make sure that young children cannot change the controls.

## Preparing Eggs

The eggs should be cleaned, labelled, dated, weighed and stored in the manner described above; they should preferably be two to ten days old. Use pencil for labelling, not felt-tips. If there is more than one setting of eggs in the incubator, mark the batches in different coloured crayons for easy identification. When eggs are turned manually, they are traditionally marked with an 'O' on one side and an 'X' on the other to check that every egg has been turned through 180 degrees.

Cool eggs should not be transferred immediately to an environment at 37–38°C, because a rapid rise in temperature can cause the yolk membrane to burst. Raise the core temperature of the eggs to room temperature at 25°C before transferring them to the incubator, which should be pre-heated for a day.

***Incubation temperatures for goose eggs compared with hen eggs***

| Species | Forced-air core temperature | | Still-air temperature level with the embryo at upper surface of egg | | Incubation period (days) |
|---|---|---|---|---|---|
| | °F | C° | F° | C° | |
| Goose | 99.0 | 37.2 | 101 | 38.4 | 28–35 |
| Hen | 99.5 | 37.5 | 102 | 38.8 | 21 |

*Figures are taken from Anderson Brown,* The Incubation Book. *It is prudent to check the accuracy of thermometers against each other and a clinical thermometer. Check for alcohol types losing liquid, or a split in the column of a mercury thermometer causing a false reading.*

## Temperature

All bird eggs typically incubate between 37 and 38°C, so an average of 37.5 is acceptable for the core temperature of forced-air incubators. Anderson Brown recommends 37.2°C as a core temperature for goose eggs; although 37.8°C has also been recommended, this feels hot.

In the still-air incubator, the temperature is 38.4°C on the thermometer level with the top of the egg. The still-air thermometer temperature at the top of the egg is higher because of the vertical temperature gradient. The temperature difference between the base of the egg and its top will be as much as 2–3°C. This temperature gradient actually mimics the situation in the natural nest where the egg is cooler at the base, but has the correct temperature at the top where the germinal disc rests against the bird. As development proceeds, the embryo grows and moves downwards from its original position on top of the yolk, eventually to fill the whole egg. Its circulation and metabolic heat then make the temperature more even throughout the egg.

It is difficult to regulate temperature closely in a small still-air incubator (forced-air models are much better). For example, if thermometers are placed around the incubator (at the same height), there can be between 1–2°C difference in temperature between the sides and the middle. However, conditions in the nest vary in the same way, in that a broody goose will exchange the cooler eggs at the edge of the nest for the warmer ones in the middle; therefore the relative positions of the eggs should be shuffled around the incubator from time to time. Alternatively, if the eggs are of different sizes, the larger eggs which stand slightly higher should be in the cooler spots, and the smaller ones in the hotter spots.

It is very important to achieve a steady, correct temperature at the start of incubation. Marginally high temperatures late on produce earlier hatches, but core temperature above 39.4°C is lethal. Unfortunately mistakes with temperature cannot be corrected by compensating the other way afterwards. Instead, try to keep as steady a temperature as possible for the remainder of the incubation period. If the eggs have been severely heated, cool them for fifteen minutes to get the core temperature back to the correct level quickly.

As incubation progresses, the embryo generates its own heat. This means that the control dial in a forced-air incubator may need to be regulated downwards to prevent over-heating, especially during the fourth week when Anderson Brown recommends a forced-air temperature of 36.9°C (98.5°F). The eggs themselves will feel hotter because they are developing the metabolism of the bird – which is several degrees above the incubation temperature.

Although the incubator is always needed to keep the eggs warm, after two weeks the egg's own metabolism raises its temperature above that of the air in the incubator. In addition, the exertion of hatching raises temperature.

### *Cooling*

There is no need to worry about the eggs and incubator cooling whilst the eggs are turned. Larger birds characteristically leave the nest to forage once each day; in geese this period varies from fifteen to sixty minutes. In a natural nest, the materials underneath and the down on top limit heat loss; however, a limited period of cooling to mimic this natural process is thought to be beneficial.

Cooling should be during the mid-incubation period,

progressing from five minutes a day from day seven, to fifteen minutes by day twenty-one. As recommended by Reginald Appleyard (*Poultry World*, Feb 1936), this procedure may improve water loss and hatchability. However, do not cool the eggs excessively, because consistently low temperatures over the incubation period slow down the growth of the embryo, and can cause problems such as delayed hatching, an external yolk sac, and deformed feet. Similar problems are also caused by high humidity, which also delays hatching.

### *Setting and Turning*

The usual incubation period for goose eggs is thirty to thirty-four days. Make sure that the likely date of pip (as early as day twenty-eight) and hatch (as late as day thirty-five for large eggs) is a period when the eggs can be supervised.

Eggs should be set on their side, not pointed end down as is the practice for duck and hen eggs in tray inserts in automatic incubators.

Eggs need not be turned for the first twenty-four hours, but after that, turning is essential. The germinal disc, where the incubated cells will divide and grow, is located on a part of the yolk which is least dense, so that when the egg is turned, the disc floats upwards. It is therefore in closest contact with the body of a sitting bird. Turning the egg not only prevents the disc and yolk sticking to the outer membrane of the egg, but also places the disc in contact with new nutrients. In the early stages, when the embryo has scarcely developed its own blood supply, it needs an immediate food source from adjacent fluids.

Eggs are generally turned each hour by waterfowl. Automatic incubators turn once an hour, and although a small turn may be acceptable in automatic incubators because turning is frequent, 90 degrees is said to be insufficient. Automatic machines of this type produce more hatchlings which pip at the wrong end. This rarely occurs in naturally incubated eggs, and hatch rates are better when turning is through 180 degrees (Artigueres Research Centre France).

Turning twice a day by hand is an absolute minimum: hand-turning three to five times through 180 degrees is more effective. Geese can turn their eggs end-over-end in the nest, though frequently they are just randomly stirred about. Turning the eggs left, then right, works well: it is important not to keep winding the egg in the same direction, because the chalazae, supporting the yolk, and the internal structure will be disrupted.

Have a routine to turning the eggs: always write down the actions, indicating the daily left/right/left turns each time, to be certain the eggs have been turned enough times. List the date of incubation and the likely date for pipping.

Hand-turning of the eggs allows frequent checks on temperature. If the thermostat or the thermometer is faulty, the incorrect temperature will be noticed by the feel of the eggs; your face is the most accurate place for testing.

Turning is unnecessary two to three days before hatching.

## Ventilation

As the embryo grows, it consumes oxygen and releases carbon dioxide: these gases are exchanged through the shell and the network of blood vessels in the allantois. Oxygen demand is low to start with, but increases to 3 pints (1.7ltr) per hour per 2oz (60g) of egg near hatching (Deeming, 2002). Embryos can tolerate 1 per cent $CO_2$ during the early stages of development and beneficial effects on hatch rates have been recorded at 2.0 per cent during the first ten days. There is progressive tolerance to higher levels (measured in the air sac) as incubation progresses.

Small incubators are designed to circulate their warm air by convection: warm air is vented at the top and cool air drawn in at the base. Therefore as long as the room temperature stays below 25°, then fresh air should circulate. Ensure that the vents are open.

## Humidity

Although temperature is a key variable in egg development and hatchability, correct humidity is crucial. The egg should lose 14–15 per cent of its original weight by evaporation of its water through the shell's pores over the incubation period. The gosling breaks into the enlarged air sac to breathe prior to hatching; this space must also give it enough room to manoeuvre to get out of the egg. Water loss from the egg is controlled by variables such as shell porosity, and the relative humidity (RH) of the incubator, which is in turn affected by the local climate. The variable which can largely be controlled by the operator is the incubator's RH.

### *Measuring Humidity*

Air contains water vapour as a gas. The amount the air will hold depends on its temperature: warm air hold more water vapour. The easiest way to measure the air's moisture is by its relative humidity (RH). The amount of water vapour the air could hold at a certain temperature (if water were freely available to saturate the air) is compared with what it is actually holding. This figure can be calculated by using a wet and dry bulb thermometer, and a conversion table based on the values.

A large difference in the two values, caused by evaporation at the wet bulb, means a low RH. For example, with a dry bulb temperature of 37.5, the wet

*Three different types of humidity meter:* left, *bimetallic;* centre, *hair hygrometer;* right, *electronic meter reading RH 35 (in late May) and 36°C (the incubator was open). Different types of meter may not agree with each other. The electronic type, which measures the change in electrical resistance of a small chip of moisture-sensitive material, is the most reliable.*

bulb reading would be only 20.6 at RH twenty. By contrast, a small difference in values occurs when there is little evaporation. With a dry bulb temperature of 37.5, the wet bulb reading would be 34.2 at RH eighty, and the two thermometer readings are identical at RH 100. When the air is nearly saturated, eggs will not, of course, lose water.

A modern electronic humidity gauge gives a quick digital read-out and allows spot checks for RH during storage and incubation.

### *Controlling Humidity in the Incubator*

The surface area of water available for evaporation affects relative humidity: if RH is too low, then filling all water trays and even spreading a damp cloth over the free surface area raises it.

Air exchange also affects humidity and evaporation rates. If there are variable vents at the base, air flow can be restricted to increase humidity, or the vents opened up to increase the draught and decrease humidity. Moving air affects RH because fresh air is constantly replacing damper air in contact with the egg. That is why forced-air incubators, which have a considerable draught, remove more water from eggs than still-air types.

Many automatic incubators now have a built-in digital humidity reader. Some of these incubators can be set for a specific RH, where water is provided, when required, by a wick from a reservoir. The water reservoir must be kept clean because warm, damp incubators are also breeding grounds for bacteria. Note that *reducing* RH is more often required for early incubation in the UK.

### *Humidity and the Incubating Egg: Examples*

Incubators always arrive with a water tray, and it seems obvious that waterfowl need high humidity for their eggs. The standard recommended humidity was RH fifty-five, and in the house, water will generally have to be added to the incubator to achieve RH fifty-five. In my experience, however, in the west of the UK, most pure-breed goose eggs cannot hatch at RH fifty-five.

**Early eggs:** The best weather for early incubator hatching is when there is a cold north wind in spring. This cool air stream often has low RH, and because it is cool, its total moisture content is also low. When the air is warmed up in the house, RH drops even further; incubator RH may drop as low as twenty on the electronic meter and Brecon eggs still have difficulty in hatching. If the incubator has variable vents at the base, opening all these for a greater flow of air to remove water vapour from the eggs can help, as can the use of a forced-air incubator. However, broodies are often essential to hatch these early eggs with less porous shells.

**Later eggs:** These eggs are often more hatchable and usually need a higher RH. This is probably because of the changing properties of the shell towards the end of the season, possibly as a result of better nutrition of the parent birds in April and May, and healthier embryos. Add water to the incubator if weight loss calculations (*see* section 'Weight Loss during Incubation' below) indicate the need, and even spray the eggs: this prevents the internal membranes from becoming desiccated and impeding the gosling from escaping from its shell. Where desiccation is too rapid, spraying and maintaining RH eighty can halt water loss; severely dehydrating eggs can even be dipped in lukewarm water.

**Porous eggs:** The eggs of Chinese and some Steinbachers are different from the Greylag types in that they lose water more easily. March and April settings can hatch in the incubator, and by mid/late season may be best run at RH fifty-five. It is therefore not possible to incubate these eggs alongside those with different properties – for example very large African eggs, Brecon and Toulouse.

### *Relative Humidity Value*

In short, there is no specific RH value for hatching goose eggs: the figure depends upon a number of

*Brinsea's OvaView candling lamp is perfect for duck eggs and also very effective for thicker-shelled goose eggs. It is battery-powered and portable for convenience. The high output, high efficiency LED illumination means that there is no overheating of the eggs during inspection.*

variables. The ideal RH for the correct weight loss varies between different breeds of geese, the time of year, and the type of incubator. There is also enormous variation from year to year even at the same location.

The most common problem in hatching goose eggs in small incubators is their failure to lose sufficient water early in the year. Large forced-air commercial incubators behave differently. They have a greater exchange of air than small models, and the draught affects RH.

### *High Humidity and 'Dead in Shell'*

Goose eggs that fail to lose enough water result in 'dead in shell' goslings – goslings that have tried to pip the membrane to breathe, but have failed due to excess liquid. This is such a problem for early eggs that it is necessary to resort to low-tech broodies. Incubating eggs in a natural nest (*see* Chapter 18, section 'Broody Hens and Muscovy Ducks') solves the problem of insufficient moisture loss in early eggs.

## DEVELOPMENT OF THE EMBRYO

By about the fourth day, the embryo has grown its extra-embryonic membranes (these are extensions of its body): the yolk sac, the amnion, the chorion and the allantois. These are the embryo's life support systems. Blood vessels linked to the heart transport nutrients from the yolk to the embryo. The yolk sac membrane is continuous with the small intestine. The allantois contains waste products from the kidneys and is eventually discarded on hatching. The amnion is a fluid-filled bag to protect the embryo. The chorion is at first surrounded by albumen, but as the albumen is used up, the chorion eventually contacts the inner shell membrane.

Between days four and seven, most of what can be seen on candling is the brain and a large rudimentary eye. This shows as a central spot within a system of radiating veins.

During the last third of incubation, the chorio-allantois allows a weak acid to dissolve some of the eggshell and thus supply calcium for bone formation.

1 Allantois
2 Amnion
3 Yolk sac
4 Chorion
5 Albumen
6 Air sac

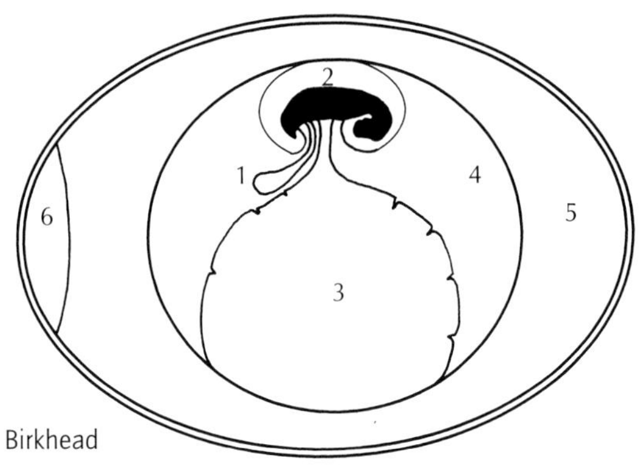


Adapted from Charnock Bradley and Brooke & Birkhead

## Candling Eggs

Candling eggs in the nest gives only limited information, and it must be done infrequently so as not to disturb the goose; a lot more can be learned in the incubator room with a bright light and a clean egg. A torch can be used for candling, but the beam may not be bright enough to shine through the goose egg. A good candling device will save work because you will be able to identify and reject infertiles and dead embryos early on.

If the eggs are clean, successful candling may be done at four to five days. A bright light is needed in a dark room; the goose eggshell is very thick in comparison with the duck's, which is very easy to candle. A fertile goose egg will have a yolk that appears to 'glow' at four days, whereas an infertile egg looks like any unincubated egg, with the yolk casting a dull shadow. If you are unsure, check again at six to seven days, when the eye appears as a definite spot and veins are visible. Eggs that have started but then died will not show an eye, and appear to slosh around in the shell as the egg is rotated. In duck eggs, degeneration of the veins after the death of the embryo causes a 'blood ring', but this is hard to identify in the goose egg. Sometimes there are black patches inside a transparent egg, but no veins; these are probably colonies of bacteria, and the egg must be removed.

At days ten to fourteen, as incubation progresses, the whole of the egg will darken and there should be a clear dividing line between the air sac and embryo. An indistinct line and a pale band adjacent to the membrane indicate that the egg has died. Dead eggs are also cooler. If you are not sure that an egg is dead, leave it in the incubator as long as it is not smelly. Eggs that have started and then died will go rotten and can explode, but if the incubator is in the house, bad eggs are noticed before this stage by their smell.

The size of the air sac should not appear to change a great deal up to eighteen days, even though there has been a fairly steady loss in weight; after that regular and more rapid change should occur visually. Mark the extent of the air sac with a pencil line to monitor development. The eggs with the best development of this feature are the early hatchers. However, if the weather is dry and the air sac too large - for example in Steinbacher and Chinese - do intervene with a higher humidity incubator before day eighteen, especially if the necessity for this change is indicated by weight loss calculations.

## Weight Loss During Incubation

Although the size of the air sac is a useful visual guide to check for correct dehydration, it is not as accurate as weight loss. Although 13 per cent weight loss has been recommended, the waterfowl egg needs to lose 14 to 15 per cent of its original weight by evaporation of water over twenty-eight to thirty days. However, for the average egg, this is best lost by day twenty-eight, or even earlier. Weight loss can also be halted in a humid incubator; it is often much more difficult to lose water.

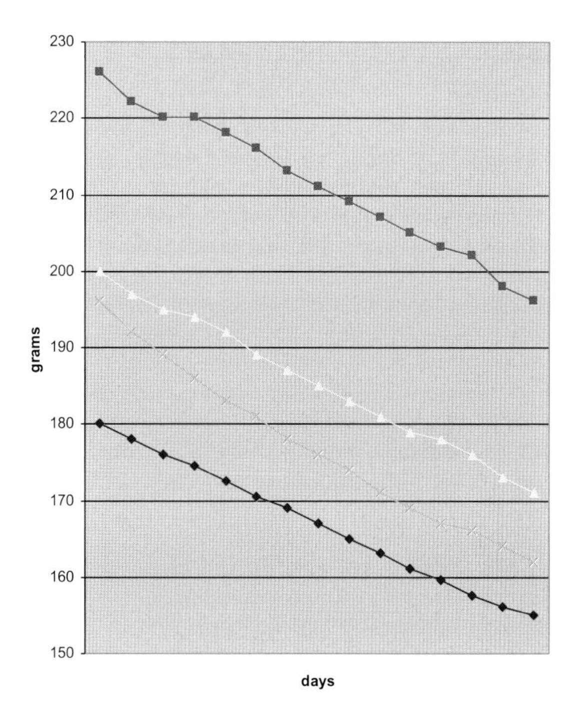


*Graph of loss in weight of goose eggs over a twenty-eight-day incubation period. Eggs will hatch between 13 and 18 per cent weight loss, but 14 to 15 per cent is ideal.*

*Weight loss in grams in sample of goose eggs during a 28-day incubation period*

| Day | 0 | 2 | 4 | 6 | 8 | 10 | 12 | 14 | 16 | 18 | 20 | 22 | 24 | 26 | 28 | % loss | hatched |
|---|---|---|---|---|---|---|---|---|---|---|---|---|---|---|---|---|---|
| Ideal | 180 | 178 | 176 | 174.5 | 172.5 | 170.5 | 169 | 167 | 165 | 163 | 161 | 159.5 | 157.5 | 156 | 155 | 14 | |
| Large African | 226 | 222 | 220 | 220 | 218 | 216 | 213 | 211 | 209 | 207 | 205 | 203 | 202 | 198 | 196 | 13 | yes |
| Pilgrim | 200 | 197 | 195 | 194 | 192 | 189 | 187 | 185 | 183 | 181 | 179 | 178 | 176 | 173 | 171 | 14.5 | yes |
| African | 196 | 192 | 189 | 186 | 183 | 181 | 178 | 176 | 174 | 171 | 169 | 167 | 166 | 164 | 162 | 17 | no |

Hatches in the UK are better if the eggs are slightly on the dry side, having lost even up to 15 per cent of their weight rather than 13 per cent.

### Method

* Weigh eggs before incubation. Digital kitchen scales are sufficiently accurate for weight loss calculations for goose eggs.
* On day fourteen, weigh again. The eggs should have lost 7 per cent of their original weight (weight loss/ original weight × 100 = 7 per cent).
* If the weight loss is correct, continue incubation under the same conditions as days one to fourteen to achieve 14 per cent loss.
* Compensate for the humidity if the eggs are either too wet or too dry. Use a broody for wet eggs, or an alternative wet incubator for dry eggs.

## Hatching in the Incubator

### Cleaning

It is preferable to keep one incubator as a hatcher, which can be thoroughly cleaned after use. It can, of course, be used as an ordinary incubator in between hatches. The pre-requisite is that it must be easy to clean (at the time of writing a Brinsea Hatchmaker is ideal). Make sure that the base of this incubator can be immersed and scrubbed with detergent; afterwards a disinfectant can be used (such as, for example, in the current market, Milton or Virkon). The interior of the lid should also be wiped down with a damp cloth and disinfectant. Incubators can be disinfected by fumigation, but the main benefit of this is in large hatcheries where there is continuous setting and hatching.

### Suitable Hatching Materials

Hatching paper is available from incubator stockists, and this helps to keep the incubator clean and the goslings comfortable. Textured paper kitchen towels can be used; newspaper is not suitable because it is too smooth – the goslings will find it difficult to stand and could end up with spraddled legs. Old, clean clothes are useful as hatching cloth; they are better than paper as long as there are no loose strands that might get tied around the goslings' legs, or eaten.

### Regulating the Temperature

The temperature of the hatcher must be watched carefully. Active hatchlings generate heat, and the incubator temperature may rise to unacceptable levels if it is not regulated. Maintain the hatcher at an air temperature of 36–37°C. Eggs that are active will feel very hot, and can be 2°C higher than their recommended core temperature because of the embryos' exertions. Those having a rest (or dead) will feel cooler. On the whole, it is better to have the hatcher on the cool side rather than too hot, which is lethal to the hatching goslings. Go by the feel of the eggs as much as the thermometer.

### Pipping

The embryo breaks out of its membrane by twitching movements of the neck, which cause the tip of the beak to jab upwards. The beak tip has an 'egg tooth' which is quite noticeable on hatching, but it then drops off.

When the egg tooth breaks the internal membrane, the bird takes its first breath into its lungs from the air sac. On breathing, the concentration of $CO_2$ in the air sac increases, and stimulates the neck muscles to twitch. The twitching helps withdraw the yolk into the body cavity, and also breaks or 'pips' the eggshell so that the bird has access to fresh air. Once this is achieved, the hatchling often rests because it needs to replenish its bloodstream with oxygen. This pause varies in duration, but during this critical time the yolk sac is fully withdrawn, and the blood vessels of the egg are shut down.

### Humidity

If the incubation temperature and humidity has been correct, the smaller eggs of the more active breeds should pip on days twenty-eight to thirty; they should then be transferred to the hatcher. Fill the water trays in the awaiting hatcher, but leave the larger, slower eggs in the main incubator until they too have pipped (days thirty to thirty-two).

If the eggs have become too dry (at any point), then halt dehydration completely by filling all the water containers in the incubator and spreading out wet cloth on the vacant floor space. However, take care in this: it is the *area* of evaporating water that controls the humidity, and it is essential to monitor the situation, because where the eggs need to be sprayed and wet cloth is used to increase the area of water, the subsequent evaporation (which uses latent heat) can adversely affect the incubator temperature and delay hatching.

High humidity is also essential after pipping to keep the membranes moist and to ensure that the goslings do not get glued to their shell. This is why forced-air incubators do not work well for hatching, because the draught of air dries out the eggs.

### Hatching

As we have seen, the gosling will rest for at least a day after pipping. It should then rotate in its shell and complete the hatching process. It is worth keeping an eye

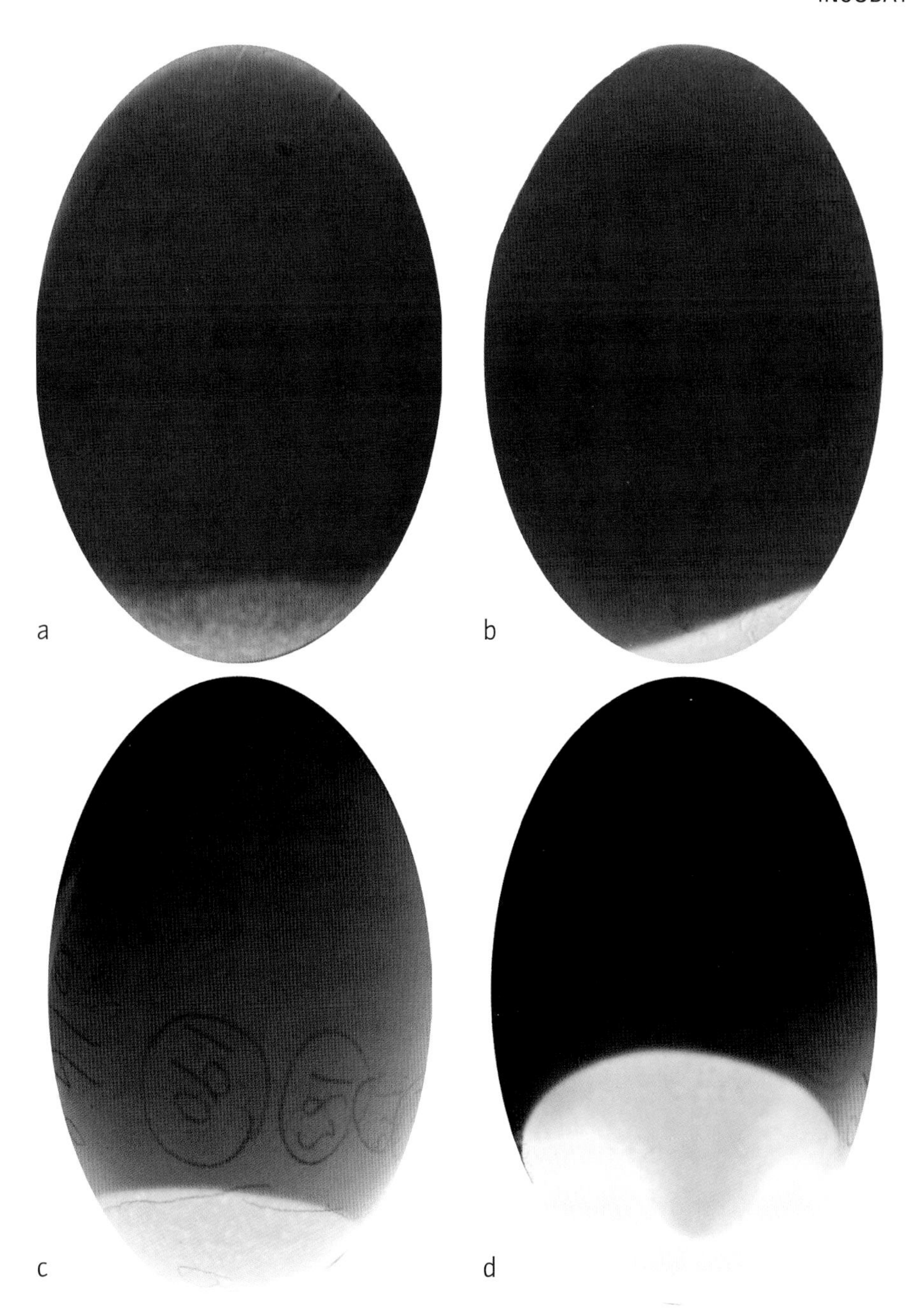


*Candling eggs during incubation – the healthy egg:*

*a) Seven days: the embryo shows as a dark spot and the radiating blood vessels, like a spider, can be seen.*

*b) Twelve to fourteen days: the egg is much darker. The embryo can scarcely be distinguished, but the blood vessels can be seen if a light is shone into the air-sac end. The margin of the air sac is clearly defined. At the opposite end of the shell, unabsorbed albumen is translucent.*

*c) Day twenty-two: the embryo has absorbed the albumen and the air sac has grown considerably. There is a rapid change in size of the air sac from this point. The change in weight is steady. This healthy egg stands a good chance of hatching.*

*d) Day twenty-eight to thirty: the egg is about to hatch; the shadow of the beak can be seen.*

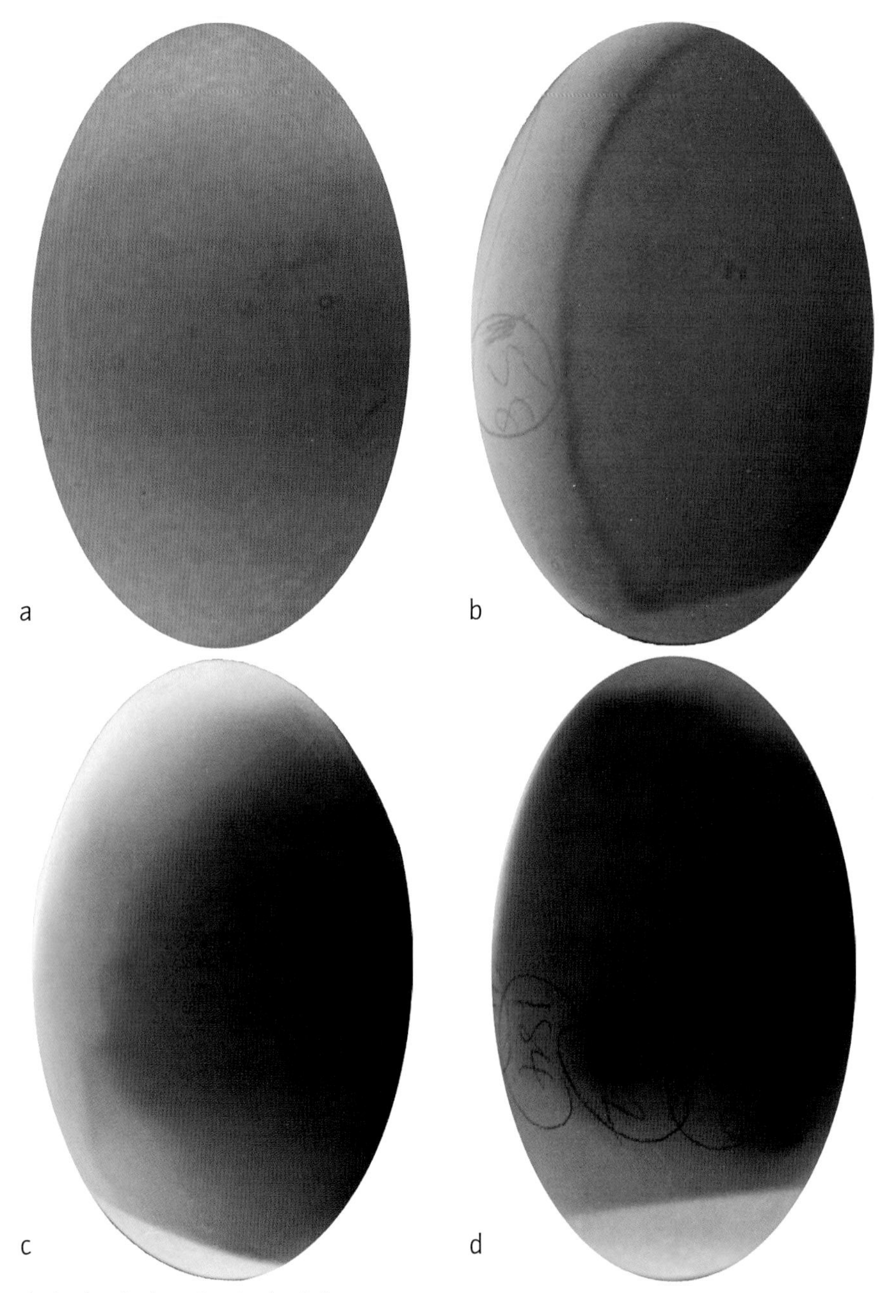


*Candling eggs during incubation – incubation failures:*

*a) Incubator clear: infertile egg with only a faint shadow of the yolk. Reject eggs like this at day five to seven. Infertile eggs do not go bad.*

*b) Fertile egg which has died: the 'blood ring' stage where the veins have degenerated. There is no eye and no network of veins. This may happen after day seven, so candle the eggs again between days twelve and fourteen. These eggs will go rotten and must be removed.*

*c) Fertile egg which has died, at eighteen to twenty days: there is a large amount of unabsorbed albumen, and no evidence of veins.*

*d) Fertile egg which has died at day twenty-two: the margin of the air sac is pale, and there are dark patches of decomposing material in the egg.*

on them at this stage, as a few unfortunate individuals then fail to flip off the lid, and suffocate. This does not happen in a nest, where the goose nibbles the eggs. If, after a day, the gosling has made no further progress, inspect the egg to see what is happening. Remove the next bit of shell: if the membrane bleeds, the bird is not ready to hatch.

Hatching can take as long as three days after pipping. It may well be that development has been retarded by too low a temperature, or the egg is too wet. Such goslings may get no further. If they die at this stage, check the dead hatchling to see if there is excessive moisture, which will run out on tearing the inner membrane. In such a case the yolk sac remains attached to the umbilicus externally, so the bird would have stood little chance of survival.

Note that oxygen demand is higher at hatching than at any other stage. Anderson Brown suggests that a hatcher needs at least twelve air exchanges per hour, compared with eight in an incubator with eggs at various stages of growth.

### Late Hatchers

Sometimes a gosling will hatch a day or two after the main hatch. Keep an eye on these stragglers. If the membrane no longer bleeds when more shell is broken away, and the inside merely looks pink, then hatching is due soon. If the membrane is dark brown and the gosling is getting smelly, hatching is overdue and the yolk sac should have been resorbed.

### Deformities

Eggs that take a long time to hatch frequently have something wrong with them. Retarded development is one cause, and deformity in the gosling is another: twisted toes or a crooked neck make hatching difficult. Deformities may be genetic, but incorrect incubation temperatures and wet eggs can cause curled toes. If such a gosling does hatch, curled toes can be corrected by using masking tape as a shoe (at the top and bottom of the web). These birds should not, however, be used as breeders, and in a large commercial hatch they would undoubtedly be humanely destroyed, if they hatched at all. Nevertheless, apparent deformities can disappear with growth if they were caused by temporary malpositioning.

### Other Problems

*Excess Albumen*

Occasionally a gosling seems to be stuck like glue to its shell and membrane. There is no excessive moisture which runs out of some 'dead in shell' eggs, but the albumen does not appear to have been properly used up. This may occur only occasionally in an otherwise normal hatch, and such goslings can be helped out with a wash in warm water to get rid of the glue and the membrane, which may stick to their eyes.

*Malpositioned Goslings*

A few goslings pip at the wrong end of the shell: they should break into the air sac at the bulbous end, but instead attempt the pointed end. They need help, because there is no air sac for them to puncture, so they may suffocate. Always clear a small air space for them to breathe, even if they bleed - and they will bleed more at this end of the shell. Nor do they have enough room to manoeuvre and perform the can-opener action of correctly positioned goslings, so they need help with this, too. Leave them alone except for checking the air supply, and only hatch them when other birds which pipped just after them have hatched themselves normally, and the membrane is no longer bloody.

To guard against malpositioned goslings, always handle eggs slowly and gently and avoid vibration and jarring. Turn eggs as frequently as possible through 180 degrees; a 90-degree automatic turn may not be sufficient. Ensure the air-sac end is uppermost.

## Summary

Hatching goose eggs in an incubator is not easy, unlike most chickens and ducks. A common experience is that very few, if any, eggs will hatch when set in early March, but that hatchability improves in May and June. The following points should be considered:

* The main reason why early goose eggs are difficult to hatch is because of insufficient water loss in the UK.
* Eggs of different breeds often need different treatment; Chinese eggs in particular and smaller African eggs hatch more easily in incubators because they lose water more readily than those from coloured European geese. Chinese eggs can become too dry in incubators and under broodies.
* The figure of 55 per cent humidity quoted for the incubation of eggs can be correct for mid- and late season duck eggs, and Chinese eggs mid-season, but is far too high for Pilgrims and Brecon Buffs. The European goose eggs need to be incubated with no additional water and very low humidity early on.
* Large eggs are more difficult to dehydrate than average eggs; large-size African eggs need 100 per cent incubation time in the broody nest, whilst average-size African eggs get too dry.

*Goslings hatch in stages in the incubator. Remove debris to make more space, and also any active hatchlings which might damage those just hatching.*

*Goslings rotate in their shell by pushing with their feet whilst they chip perforations in a circle with their egg tooth. They then push off the cap of shell. The goose assists with this in the nest; periodically check that the hatchlings are not stuck.*

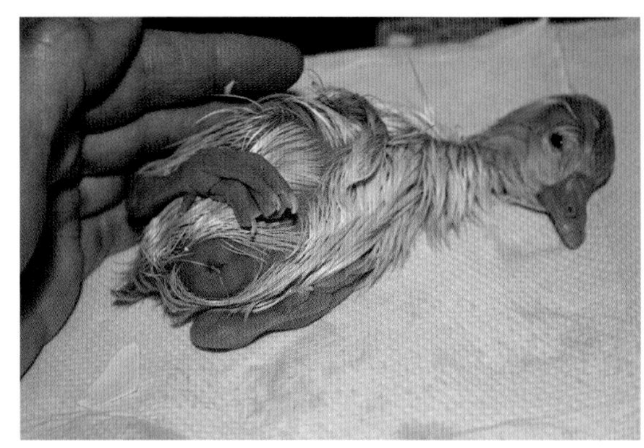

*Water loss has been correct, so the umbilical cord has broken off and left a clean umbilicus with no blood.*

* Later eggs are much more successful at dehydrating than earlier ones, despite the ambient humidity being the same or higher; this is probably because later eggs have a more porous shell, allowing easier dehydration.
* Early goose eggs hatch best under broodies, which usually get everything right, whatever the weather. The environment of the nest (*see* Chapter 18, 'Microbiology and the Nest') is crucial.
* The incubator will be a better option than broodies later on in the season, probably because it is cleaner than the nest, and humidity can be regulated; eggs under broodies can lose too much water later in the season.
* Weighing the eggs and comparing their weight with an ideal weight-loss chart is the only reliable way of charting an egg's progress and making correct adjustments to humidity.

# 20 Rearing Goslings

## Introduction

Pictures of geese out with their goslings look idyllic, but disguise the actuality, which is that mortality can be very high. Konrad Lorenz cites high losses of Greylag goslings by two weeks of age by factors such as predation and parasitism. Geese do not produce a lot of offspring each year, and every gosling is precious, so it is best to guard them very carefully until they are three to four weeks of age. During this time the gizzard and alimentary tract grow so rapidly that the gosling is largely a gizzard on legs, and it is very important to protect them during this crucial stage.

Rearing goslings is not an easy task: all sorts of different problems can impede their development to adulthood, and the following checklist to ensure relatively trouble-free progress might be helpful to the less experienced producer:

* Get the right food. Non-medicated crumbs and pellets made for growing waterfowl are best.
* Provide appropriate bedding: non-slip surfaces for the newly hatched, with dry, absorbent bedding and enough space to keep them clean.
* Provide plenty of clean water for them to wash their faces, but make sure they can't drown.
* Goslings must have clean grass, but they cannot be reared on grass alone. If they are well fed in the first few weeks, they get a good start and grow into bigger, stronger birds.
* A big advantage of rearing goslings indoors and then on pristine grass is that they remain parasite-free for several weeks.
* Goslings are very fragile to start with, and must be kept warm and dry. Protect them from rain until they are fully feathered on their backs.
* Make sure there are adequate facilities for housing them in a controlled environment. A rearing shed and grass run are essential for protection from extreme weather conditions and predation from rats, stoats, cats, dogs, crows.
* The goslings can be entrusted to a tame foster goose and gander when conditions are such that they can be let loose; alternatively they can just be kept in their hatching group.

Handled quietly, goslings are a delight to rear. They are very tame and easy to manage. Talk to them and greet them to keep their confidence.

## The First Three Weeks

### Diet and Food Quality

For a few goslings, a mixture of clipped grass, chopped dandelions and not more than 20 per cent cooked egg yolk is a good emergency starter diet - but their appetite soon grows, and it is better to feed waterfowl starter crumbs. The protein content of the crumbs is around 18–19 per cent; some brands contain fishmeal which is thought to be beneficial, adding methionine to the diet. Introduce the larger grower pellet by two to three weeks so that goslings can choose the size they want. The lower protein content of 15 per cent will help to avoid leg and wing problems.

In both crumbs and pellets, it is preferable to choose a brand manufactured for ducks and geese. *Chick* starter crumbs may not contain sufficient vitamin B, and often contain a coccidiostat which has been tested for poultry but not necessarily for waterfowl. Waterfowl should not need this additive. Ethopabate as an aid to control in coccidiosis does seem to be all right, but other additives may not be safe.

Also, check the food preservative. Preservatives are used in animal and human foods to prevent the growth of harmful moulds. Butylated hydroxyanisole and ethoxyquin can be added to animal feed, but there is concern over their effect. For short-lived, commercial animals, problems may not emerge with the small quantities involved. However, in long-lived animals and birds, chronic degenerative liver disease and arthritis are alleged to occur. Certainly birds that are fed a more natural diet - mostly grass - do live longer.

Propionic (propanoic) acid is also used as a preservative to inhibit moulds and bacteria. Used extensively in human foods, it occurs naturally as a product of digestion in herbivores, and seems to have a better track record. Do check labels carefully before buying feed, and with the company nutritionist if necessary.

Trace minerals such as selenium, copper and magnesium, plus vitamins, are added to grower pellets.

*Dandelion leaves are good for goslings, and having greens to eat stops them from 'grazing' each other. This can become a serious problem in Brecons and Pilgrims: the birds must have something to pull otherwise they will pull each other's fluff, even until they raise blood.*

Nevertheless, additional vitamin B is beneficial: brewers' yeast is a good source, and can be sprinkled on slightly moistened food or put in the drinking water (one teaspoon per litre).

Goslings kept indoors should also be given plenty of greens. A clean patch of lawn for the express purpose of freshly cut clippings is useful. Also feed dandelion leaves in a bunch as a treat each day, and graduate to young, chopped goose grass and sow thistle. An additional benefit, especially if they can pull at greens, is that goslings do not start to pull each other's fluff. Deficiency in the diet is unlikely to develop if the birds are outside and on greens early enough.

## The First Food and Drink

Exhausted goslings rest for twenty-four hours and do not need food for up to forty-eight hours after hatching, as they are still using the nourishment from their internal yolk sac. However, offer them water on the first day and food on the second. A cylindrical egg cup, too small to drown in, is a good drinker. If they seem hot or dehydrated, and to get them drinking, offer water on a teaspoon.

If a few birds are not keen to start on crumbs, they can have their appetite stimulated with greens. The best for this are tender dandelion leaves which can be finely clipped (not near the bird) with scissors, or tied securely in a bunch for the goslings to pull at.

## Temperature and Surfaces

Get the goslings out of the incubator as soon as they have fluffed up. The umbilicus can be a bit raw, so dry this off quickly (remove them from a dry incubator) to prevent infection. Transfer the goslings to a pre-warmed box lined with crumpled old clothing (cover them with it to start with), towels or chopped straw, not newspaper; place the box beneath a heat lamp (*see* below).

Do not overheat the goslings; there must be a cooler area in their box so they can escape if the height of the lamp is incorrect. Watch their behaviour – they pant if they are too hot. When they are first hatched they do not have the energy to move far, nor the knowledge that there might be a better position. Keep an eye on them until they are mobile. When they are active, their behaviour will indicate if the temperature is right: they will huddle together under the lamp if they are cold, and will spread out, away from the lamp, if it is too hot.

Prolonged exposure to heat lamps is particularly bad for goslings: they grow too quickly, and develop soft bones. Weak hocks can result, and toes start to curl, rather like poultry overexposed to infra-red lamps, and indicative of vitamin B and D deficiency. Place the food and drink away from the lamp to avoid splashes on the bulb and also to keep the food cool; vitamins deteriorate rapidly in direct heat.

It is best to get the goslings out on clean grass and exposed to some daylight after ten days – though sometimes a cold wet spring makes this impossible. If goslings are kept indoors for three weeks, then they must have greens and brewers' yeast added to their diet.

## HEAT SOURCE

When the young goslings are brought out of the hatcher they should have a heat lamp suspended above them so that heat is directed down on to their backs. The heat-emitting glass bulb (clear or ruby) is the easiest to obtain, but is not recommended: the glass can shatter, the element can blow quite unpredictably, and the twenty-four-hour feed regime that it promotes is not beneficial for goslings. The ceramic bulb (which does not emit light) is more expensive, but it lasts longer and gives the goslings a quiet night. If they have started to pull at each other's fluff, then the darkness is a big advantage. The bulbs are made at 75, 150 and 250W, so choose one appropriate for the circumstances - namely the size of the group and the ambient temperature. Ensure the lamp is suspended safely, and is not a fire hazard; if it falls on the goslings with the power still on they will be killed by the heat if not the weight.

*The ceramic bulb does not emit light. Ensure the lamp is suspended safely.*

***Regulating temperature over the first three weeks***

| *Floor temperature* | *Degrees Centigrade* | *Degrees Fahrenheit* |
|---|---|---|
| Day One | 36 | 96.8 |
| First week | 30–32 | 86–89.6 |
| Second week | 23–25 | 73.4–77 |
| Third week | 20–22 | 68–71.6 |

*The floor temperature is regulated by the height of the lamp above the floor. Note that the temperature on the gosling's back will be slightly higher, and there should also be cooler floor space for them to move into, and no draughts. Raise the lamp every few days. Heat is not required from the end of the third week as long as the ambient temperature is around 20°C.*

### Early Familiarity

Many hobby breeders like to rear the goslings in the house initially, for security and interest. It is advantageous to keep a close eye on the goslings during the first few days, and for them to have an eye on you; it is also useful to accustom them to being handled - then later they will consider this a normal activity when they are picked up to have their weight checked; and when they are adults they are easier to handle in the breeding season, and when caught for checking health.

When approaching birds that cannot see you - for example if they are in a coop or a shed - speak to them first to let them know you are there: this avoids a panic when the door is opened. Always move slowly, to give them confidence, and talk to them. Geese are great communicators, and goslings in particular need to greet each other and their handler to feel secure.

Early familiarity with people produces calm birds, and in this regard it is also of economic importance that birds in a large flock are not stressed by their handlers: if they are, they will tend to panic and crowd into a corner, and in the press risk suffocating each other. Early exposure to the human voice and to people moving about prevents this.

*Three-week-old goslings out on clean grass. These hand-reared birds follow their keeper, but they have also imprinted on each other. They get used to being left in a group, and will behave normally as adults.*

## Early Problems

### *Mortality Rate*

Close observation of goslings allows frequent health checks. However hygienic the eggs and the incubator, some birds will become ill. It is possible to hatch 100 ducks and to lose less than 2 per cent in the first two weeks after hatching, but with geese the mortality rate can be higher, depending on how easily the goslings hatched. In a healthy hatch all will live. Where there are 'dead-in-shells' and late hatchers, two to three out of fifteen may die. These losses can be reduced by sterilizing the incubators thoroughly, watching the birds carefully, and treating them appropriately.

### *Failure to Eat*

Pick up and examine any gosling that fails to eat, or is slow. Poor doers will have thin legs, will start to drop their wings, and have little energy to move. Isolate any such gosling, which may have an infection - though keep another one with it for company. Treat them both with soluble antibiotic powder (from the vet) dissolved in the drinking water, for six to eight days. Also persuade the gosling to eat: try finger-feeding. The medicated liquid can be mixed with the crumbs into a porridge, with additional brewers' yeast. This enhances the flavour and encourages the bird to eat.

### *Single Gosling*

If one gosling hatches, it will need a companion if there is no goose to look after it. Goslings need something on which to imprint. It will cry if it is left on its own, fret and not eat properly, and will imprint solely on its keeper. To prevent this, find another bird of a similar age. Another gosling is best, but failing that, ducklings are useful. Eventually try to join up the gosling with another group of younger or older goslings - but this must be done with care because the group birds may not accept the newcomer.

### *Spraddled Legs*

Sometimes large, soft goslings that have retained too much moisture are too heavy to support themselves, and their legs may splay out to the side. They should not be encouraged to eat because losing a little weight may help. Weaker goslings of yearling geese are also more prone to this condition.

It is important not to let these spraddlers dehydrate if they are unable to get to a drinker, but offer them water from an egg-cup or on a spoon.

These goslings need a rough surface to get a good grip: a crumpled old woolly jumper is the best, and straw is also good. In a bad case try hobbling the gosling's legs together, with double-strand soft wool around the ankles. A rubber band, knotted in two places, is even better, the two outer loops being stretched over each foot. Ensure it is not too tight, and remove it as soon as possible.

A bad case of spraddles may not be able to stand up, and unless helped, will dislocate its hip joint and have to be put down. Supervised remedial swimming in warm water also helps, if you have the time and patience.

*Preventing spraddled legs: the knotted elastic loop is stretched over the toes, and fits snugly round the shank.*

## Transferring Goslings to an Outdoor Shed

After a few days indoors, depending on your level of interest and the family's tolerance, the goslings can be moved to a rat-proof shed with a reliable electricity supply. The outhouse floor must be well warmed first by the heat from the heat lamp.

### *Heat Source*

It is safer to have two heat sources in case one breaks down; the lamps can be shared between two groups separated by low partitions. The partition also functions as a draught excluder beside a warm spot.

When the goslings are small they can be kept in groups of up to eleven. If more than this huddle under one heat source, the goslings in the middle risk being sat on by the rest and suffocated. If the shed is at all draughty, a three-sided cardboard box (with the flaps and top cut off) makes a nice warm spot beside or under the heat lamp. There is no 'roof' to the walk-in box, so the cardboard should not present a fire risk. Where there are larger numbers of goslings, the ambient temperature must be maintained at a comfortable level; larger, gas heating appliances are often used on a commercial scale.

The heat source, suspended from the ceiling, is gradually raised as the birds grow. The aim should be to 'harden them off' before they go out. Gauge if the lamp is the correct height, from their behaviour. Note that overnight temperatures may be significantly lower than daytime ones.

Ensure that the goslings cannot chew any electric wire or touch the heat lamp, and that there is no risk of fire from materials being too close to the heat source.

### *The Shed Floor, Litter and Space*

The floor of the shed should be concrete, to prevent rats entering, otherwise goslings can disappear in the night. Mice should also be excluded because they will contaminate food. The floor should be thoroughly cleaned and dried before the goslings are transferred to the shed. Washing soda or Virkon are useful cleaning agents.

White wood shavings can be used as litter for young birds, but are unsatisfactory to start with because the goslings will eat them. Chopped barley straw is good because the smaller pieces are easier to handle than full stalks, and the length that can be eaten is limited. The goslings will pick at the wisps of leafy material which come off the stem, but this does not seem to matter. Do not attempt to pull out long lengths of straw from the beak: it is best left where it is and just broken off. If coarse builders' sand or concreting sand is supplied, any roughage will be broken down in the gizzard.

*Chopped barley-straw litter: the pieces are short in case some are eaten. The green tray catches spillage from the water bowl. Partitions are made from interlocking plywood. The stocking rate here is eight birds per square metre.*

The floor litter can be topped up with more chopped straw, and then shavings introduced. Given time, the goslings no longer eat the bedding. Gauge when the bedding needs to be completely removed by smelling (at floor level) the amount of ammonia that builds up under the surface. Harmful moulds will develop and the bedding will need replacing with a new layer of just shavings.

Partitions to separate groups of goslings need only be about 12in (30cm) high. Hardboard bent into a circle is frequently used. If floor space is limited, squares are more economical: use interlocking pieces of thin plywood which can be scrubbed down and reused between batches, and from year to year. It is important to make sure that a gosling cannot get its head and neck stuck in any joins.

For commercial birds, a maximum density of fourteen to twenty birds per square metre during the first week, and seven to fourteen birds during the second week, is suggested. However, this is insufficient space for birds to exercise themselves, it will create dirty conditions, and will lead to them pulling and chewing at each other's downy feathers. Half those rates are preferable, namely seven and three to four respectively.

### *Water Supply*

It is safest at first to provide water in a drinking fountain, but because the rim is near floor level, the water is easily fouled. A deeper water container works better for goslings; they need to wash their eyes, and a heavy

*These Africans are two weeks old. Having the water container on weldmesh, over the drip tray, keeps the bedding (now shavings) dry. Sand is provided in the water container. The stocking rate here is two to three birds per square metre.*

pie dish or casserole, which cannot be turned over, is ideal. Goslings should be able to get out of it if they choose to get in. A stone can be placed in the middle to dissuade them from bathing and getting soaked. A dish also functions as a foot bath because the goslings inevitably get in it from time to time, and this keeps their feet in good condition.

A further advantage of young goslings being able to dip their heads in water is that they learn to condition their fluff and feathers by using the preen gland. Well oiled, preened feathers repel the rain better, and the gosling is less likely to get waterlogged.

## Coops and Runs

On warm days, the goslings can be transferred to a coop and grass run where mesh will protect them from cats, crows and magpies. Make sure that they cannot escape where the ground is uneven; block any hole with a brick. Put plenty of bedding on top of a paper sack liner in the coop (remove all string), and show the goslings this warmer area. They are slower to realize this than ducks.

If the weather is cool and wet, cover the run with polythene, or leave the birds in the shed for longer. If the weather remains poor, put them out in the covered run for a short time each day. Several goslings will huddle up and keep each other warm when they are not grazing; if there are only two or three they cannot be left out for long. Goslings of this young age must be kept warm and dry, and in bad weather they will need to go back to the heated shed overnight.

It is important to keep the backs of downy goslings dry: they will catch pneumonia if they are left wet and cold for prolonged periods, and need protection from heavy rain. Goslings that have been caught in heavy rain or hailstones, or have got stuck in a drinker and seem nearly dead, should be blotted dry immediately and put under a heat lamp to dry and fluff up. Even birds which seem about to expire can surprisingly pick up again.

The grass for the runs must be short and free from bird droppings. If possible, rear goslings on ground which is otherwise unused by geese, as this will reduce the incidence of gizzard worm. The garden lawn is ideal for this short period, and the goslings will improve it.

*A coop and grass run: the wire top (removed here) provides protection from aerial predation for two- to three-week-old goslings. A polythene cover and solid sides give shelter from wind and rain.*

However, because they consume enormous amounts of grass – and also deposit copious amounts of droppings – the runs will need moving on three to four times a day by the time they are three weeks old.

## Rearing Downy Goslings with the Goose

The first three weeks with a goose require close observation. One good system has the goslings where they can be seen by the parents inside the coop; they are later allowed to join them. In this way, the geese are familiarized with the offspring and well bonded. Generally this can only be done with a goose which has hatched and reared goslings before; a yearling goose must hatch something first. Geese unfamiliar with goslings, and especially females, can chivvy and even attack them. On the other hand, some ganders make excellent fathers. They are more careful with their feet than the females, and less likely to tread on the young; some take guard duty very seriously and will stand with their head up, on watch, all day. Showing a behavioural trait noted by Harrison Weir (and in wild geese), ganders will summon their goslings by paddling their feet up and down on the spot (*see* Chapter 6).

Initially, a goose and newly hatched goslings should be confined to a rat-proof, protected area which is under cover. Do not allow the goose to march off over the whole farm with young goslings: they will get lost in the grass, stuck in a ditch, fall foul of some machinery or be devoured by hawks and crows. They need to have an obstacle-free, confined, safe area.

Most medium and light breeds of geese brood their goslings well and will protect them from showers, but they are not very good at protecting them from predation. If magpies and crows are wary of your area because it is well keepered, then you might let the goose have two- to three-week-old goslings by the house, if you can watch them. However, it is best to keep goose and goslings indoors in a large, clean shed on starter crumbs for a week or more until they are all well on their feet.

Quite apart from predation, there are other problems in letting the geese have the goslings back after the late ones have hatched in the incubator. Firstly, the adults will eat all the food. Goslings out on grass will not grow very quickly without crumbs, and they need to grow fast so that predators will take less interest in them. One way to stop adults taking the crumbs is to make a gosling creep: a circle of pig wire for the goslings to pass through to get the food is usually effective. This must be large enough for the food and water bowls to be placed so that the adults cannot reach them.

*Week-old Pilgrim goslings out with their parents. Attractive as this may seem, some of these goslings were lost to falcons in their second week. Buzzards and corvids will also take them. Keep goslings very safe until aerial predation is no longer a threat, usually after four weeks of age*

Secondly, the adults must be wormed (if this has not been done already when the goose sat). More goslings are lost because their keepers are ignorant of gizzard worm and the damage it does than any other cause.

Thirdly, the water containers are a problem: a bucket is quite all right for adult geese because they are too big to upend in it, but goslings must be started off with small containers such as water fountains, and graduate to duck drinkers and washing-up bowls. Bowls should be anchored down with a brick to stop the receptacle turning over, and to stop a gosling getting stuck on its back (and drowning) in the bowl. Containers with a smooth surface, such as a bath or ceramic sink sunk into the ground in the garden, are a death trap for goslings, which cannot get a grip on the smooth surface to get out. If they fall in, they become waterlogged and drown.

Lastly, different family groups of goslings do not mix well. The larger, stronger goslings will be thoroughly nasty to smaller goslings, and because of this the family groups are best kept separate.

## Rearing After Three Weeks

After three to four weeks the goslings are more robust, and depending on the weather and the size of the breed, they can be put out in paddocks on their own or with trusted, tame geese. The smaller breeds should be over four weeks old before they are left outside all day. The goslings should not be grazed with larger animals:

inquisitive lambs, bullocks or horses can chase them and would risk trampling them. Never put geese, at any age, into the same confined area when herding sheep and cattle.

There are two different routes to take for rearing the goslings after this point, depending on the amount of space and pasture: they can be reared mainly on grass, or mainly on pellets and wheat. Each method has its own benefits and problems, and in practice geese are best reared with access to grazing in the daytime, and some supplementary feed. The grass provides the essential vitamins that the pellets may lack; the pellets provide the added minerals and extra protein needed for growth in the heavier breeds and breeder-quality geese.

If the cost of production is crucial and the birds are destined for the table, then rearing on grass alone with fattening feed given at a later date is the most cost effective.

## Supplementary Feeds

At three weeks the goslings should be graduating to growers' pellets, which are 15 per cent protein; by five weeks it is wise to start adding just a few grains of wheat to the diet. In no circumstances should they continue on just starter crumbs: not only is this wasteful, as the birds will scatter the crumbs, but it also makes the condition known as 'dropped tongue' (*see* Chapter 22) more likely. Also, the protein content of the crumbs is too high.

Goslings do not particularly like wheat at this stage, but the protein in the diet must be cut down over the next few weeks. The diet has to be watched particularly carefully where birds are reared with little access to grass. Where they have plenty of grass, sunlight and exercise there are fewer problems regarding too much protein, vitamin deficiencies, or weak legs and wings.

Whilst the goslings were on deep litter they needed access to sand for the gizzard to break down any bedding ingested. Although pelleted food is easy to absorb, sand should always be available for the health of the gizzard. Once hard grains and grass are fed, sand and grit are essential for the gosling to make use of these foods.

### *Type and Amount of Food*

Poultry grower pellets frequently contain the same coccidiostat as chick crumbs. This is not a treatment for the disease if the birds become ill, but helps prevent it occurring. Rations containing coccidiostat are generally not recommended for waterfowl, so duck/goose grower pellets should be sought. If waterfowl food is not available, fattener pellets containing no such additive could be used. However, the protein content must be cut by mixing pellets with wheat.

Layer pellets are designed for laying chickens and may contain 4 per cent calcium for shell formation, as well as undesirable additives such as egg yolk colouring. They are not suitable because the excess calcium will cause problems (*see* Chapter 16).

***Duck/goose starter and grower rations compared with a layer ration***

| | *Starter* | *Grower* | *Poultry layer* |
|---|---|---|---|
| Oil | 5.5% | 4.5% | 3.5% |
| Protein | 18.5% | 15% | 16% |
| Fibre | 4% | 4.5% | 7% |
| Ash | 6% | 7% | 15% |
| Methionine | 0.3% | 0.25% | 0.35% |
| Moisture | 13.8% | 13.8% | 13.80% |
| Vit A | 15000iu/kg | 8000iu/kg | 8000iu/kg |
| Vit D3 | 3000iu/kg | 2000iu/kg | 2000iu/kg |
| Vit E | 80iu/kg | 28iu/kg | 10 iu/kg |
| Sodium selenite: selenium | 0.31mg/kg | 0.31mg/kg | not specified |
| Copper sulphate: copper | 25mg/kg | 25mg/kg | not specified |

*The rations are made from wheat, soya, peas and linseed, which provide natural methionine and lysine. They also contain calcium carbonate, di-calcium phosphate, and salt. Layer pellets do not have the same vitamin and mineral content as grower pellets. Always store food in cool, dry conditions to avoid the growth of moulds, and to preserve vitamins.*

Goslings with no access to grass will eat enormous amounts of food. For example, between eight and sixteen weeks growing Embdens will eat at least 1lb (450g) and Brecon Buffs up to ¾lb (350g) of mixed wheat and pellets per day. This compares with 4–8oz (112–225g) for most adult stock birds (depending on the time of year). Considering that commercial birds are fed until Christmas, it is not economic to produce table geese under a zero-grazing regime.

### *Wing Problems*

Intensively fed goslings must be watched carefully at four to five weeks when the blood quills sprout. If growth is very rapid, the weight of the quills cannot be supported by the muscles and bones of the wing. This results in the following problems:

**Dropped wing:** The quills are too heavy for the gosling to carry the wing in the normal wing position. However,

*Dropped wing: this eight-week-old large Brecon gander has grown very rapidly, but his muscles cannot support the weight of the blood quills. The wing was taped up and set in the right position, and was correct at fourteen weeks; exercise and swimming water helped.*

it may pick up the wing later and be all right. This condition can be helped by taping up the wing to take the weight, and setting it in the correct position. Slower-growing females on the same food regime are not affected.

**Rough wing:** The primaries are not folded neatly in their normal position but sit slightly turned out.

**Slipped/oar/angel wing:** The final joint is turned outwards. This occurs in domestic birds which have access to large amounts of food. It would be lethal in a wild population because such a bird cannot fly. In a few cases this can be an inherited weakness.

It is therefore important to accustom the goslings to a few grains of wheat in the diet from four to five weeks, so as to cut down on the protein. If there is a mixed diet of grass and bag feed, then underfeed the pellet/wheat mixture at this stage, until the feathers are safely sprouted and set in place by seven to eight weeks. The wheat/pellet mix should be 1:1 at this stage. Once the feathers are fully formed and set, more wheat/pellets can be fed to get size in the larger breeds. Slipped wing is less likely to happen in grass-fed goslings because their rate of growth is slower.

*Cultivate a patch of clean, tender grass for young goslings. It should be the main part of their diet.*

## Rearing Mainly on Grass

From three to four weeks the goslings can be reared on much less supplementary food, depending on the breed and conditions. Even on free choice of food (grass and bag feed), consumption of the wheat/pellet mixture is generally lowered by 50 per cent.

If they are to be grass reared, then the stocking density of the birds must be low. They should be grazed after other animals such as sheep and cattle, which keep grass short. Succulent spring and early summer grass is high in protein, but this drops by mid to late summer, and when grass is longer. Geese, and goslings in particular, do not thrive on long, tough herbage: they like to strip the seed heads off mature grasses, but growth is achieved by eating young shoots (*see* Chapter 6).

If the grass is coarse it can cause impaction in the proventiculus. Unlike the chicken, geese do not have

*Development of the feathers:*
*Week 3–4: Blood quills start to grow from the end of the wing. Scapular feathers have grown, but the back is only covered by downy feathers.*

*Development of the feathers:*
*Week 4–5: Blood fills the hollow quill, and feathers sprout. The wings are very heavy at this stage, and the bone can be pulled outwards and permanently distorted into slipped wing.*

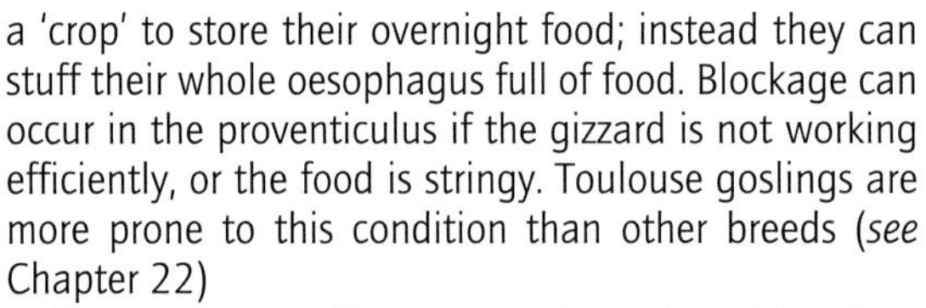

a 'crop' to store their overnight food; instead they can stuff their whole oesophagus full of food. Blockage can occur in the proventiculus if the gizzard is not working efficiently, or the food is stringy. Toulouse goslings are more prone to this condition than other breeds (*see* Chapter 22)

Even on top quality grass, goslings should be given supplementary wheat and pellets, with free access to sand and water. This food mixture is best given in the early evening, to save wastage and to encourage the geese to range for grazing in the day. The amount fed will depend on the cost of food, the state of the grass and the health of the birds (*see* Chapter 21). There is no need to leave food and water in the shed overnight; the water will end up making the bedding a soggy mess.

*Development of the feathers:*
*Week 7: Primary feathers are well grown, and correctly set. Scapular and back feathers also protect the back from rain.*

## Common Internal Parasites in Grass-reared Birds

If grass is the main part of the diet it must be clean and free from parasites. Breeders who stock ducks and geese intensively only manage to stay fairly free from trouble caused by parasites through good management and feeding most of the diet from bags. Parasites causing the most problems in grass-reared birds are gizzard worm, also gape worms and coccidia (*see* Chapter 22). These problems can also arise in birds fed on concentrates with access to some grass, and these diseases should be borne in mind if birds seem ill.

Infestation by gizzard worm is one of the most common causes of gosling deaths. Its incidence would be much reduced if all geese were wormed strategically when sold and moved on to a new property. Once the disease is recognized, it is too late; the parasite is established. Avoid problems by grazing geese on clean pasture at a low stocking density so that they do not pick up more worm eggs from droppings on the grass. The goslings should also be picked up regularly to see if they are heavy. If they seem light, worm them as a precaution.

Infestation at an advanced stage causes a change in behaviour. If goslings are listless and easy to catch because they have no energy, they are very ill – in fact they are starving to death because the gizzard cannot function to prepare their food. Goslings that have

*This 'duck drinker' is an excellent design for goslings. The water stays clean, and they cannot drown. They also need a large, wider container to throw water over themselves and to wash their faces.*

*This shallow footbath designed for sheep was successfully used for rearing this Brecon flock. It is fed by hosepipe and drains into a natural ditch. Chris Lea*

deteriorated this far are difficult to save because they lose their appetite. They can sometimes be rescued by liquid feeding. The proprietory instant drink powder for humans known as Horlicks, made into drink form, is suitable for thin goslings because it contains milk and other nutrients; mixed up as a concentrated liquid and administered carefully down the throat with several refills of a 2ml syringe, it can revive the appetite.

It is better to avoid this eventuality by regular inspections and worming with Flubenvet if neccessary. The gizzard worm cycle is as little as three weeks, so if goslings begin to graze at three weeks, worm them at six to eight weeks if they are light - that is, if necessary. If they are scarcely eating, or a rapid response is needed, a liquid oral, single-drench vermifuge can be used (*see* Chapter 22).

### Water Supply

At every stage, goslings must have a supply of clean water. This requires that, unless an automatic water supply is available, the drinkers are replenished several times a day, especially in hot weather. A combination of containers is often useful: a bell drinker which is generally not fouled by droppings, and a larger container for bathing.

## Sexing Goslings and Adult Geese

### Colour Differences – Auto-sexing

White breeds such as Embden, Roman and Sebastopol are auto-sexing as infants, but this is not 100 per cent reliable. Both males and females show a faint grey-back pattern in the fluff at hatch, but this fluff is darker in the females, which can look like Grey-Backs at this stage. These markings are replaced by white feathers, but females may retain some grey on the rump until their second year.

Auto-sexing breeds such as West of England, Shetland and Normandy also show the grey-back (pied) pattern in the fluff even more clearly in the females, and of course retain this grey-back pattern in adult plumage. Pilgrim geese likewise hatch all-grey females, which retain grey feathers. Auto-sexing female goslings have a darker beak than the males, which have pale beaks and yellowish-grey fluff when first hatched.

The reliability of these auto-sexing breeds does depend upon the purity of the breeding pair and the knowledge of the breeder. So-called 'Pilgrim' geese can produce white females, and so-called 'West of England' grey and white geese turn out to be male; therefore birds may need to be vent-sexed to check if they are breeding true.

In breeds that lack the dilution gene there is no sex

difference in the colour of the feathers or the bill – for example Brecon Buff, Grey-Back, Steinbacher and Toulouse. In the case of such breeds, other indicators need to be used.

### Size and Shape

Birds can be sexed by size and behaviour. Male goslings are faster growers, and for a given strain and hatch of geese, larger birds are generally male. This is found on close-ringing, in that males have bigger feet. The size difference is marked in Africans and Chinese, where the males are distinctly taller at sixteen to twenty weeks. Eventually the knob on the head of these breeds grows larger and broader in the male, but this does take time.

Ganders' heads are usually bolder than females', and their body carriage higher. The goose should have a heavier undercarriage even by twenty-six weeks, though well fed Brecon and Embden ganders also have a dual-lobed undercarriage.

### Behaviour

Spending time with adult birds allows the observation of differences in behaviour which indicate the sex (*see* Chapter 17). Young birds should be handled regularly to check their weight, and to keep them tame. Male goslings tend to be more confident in coming forwards, although one has to allow for differences between the breeds; thus Toulouse are shy and Africans are more forward, but for a given group of birds of the same breed, the males are generally more forward than the females and easier to pick up.

Confidence in the males also shows when they are held facing you. Tame females will chew clothes and your hair more than males, which sit more calmly in the hands. As a general rule, it works for all the breeds.

Incidentally, goslings do fight – even in the first few weeks they will scrap, instinctively using adult behaviour, one bird grasping another by the neck and beating with outstretched, stubby wings. This has usually happened before they are sexed, but has been noted in African males at six weeks.

### Voice

From around twelve to sixteen weeks of age, the female gosling voice deepens; the gander's voice remains higher, and is more rapid in delivery. Listen to the goslings' chatter carefully: it is a useful confirmation of their sex.

This difference in tone is an adult characteristic. Brecon females have a low, gravelly voice, whilst the males have a higher, more rapid chatter. It is also the adult male who stretches up and gives a really high-pitched shriek.

Chinese and Africans are different: Chinese females develop a characteristic 'oink' by sixteen weeks. Africans are more difficult to distinguish by voice until older, when the females will begin to sound more like their Chinese sisters.

### Vent Sexing

It is much easier to sex young goslings than adults; although this can be done in the first week, it is nevertheless preferable to leave it until the birds are three to four weeks old. At this age the goslings have hardly sprouted quills (which would get in the way), and are larger; also they are still tame and easy to handle, so are less likely to struggle and injure themselves. It is easier to see what is there in these larger goslings (do wear glasses, if necessary, and carry out the inspection in good light), and there is less likelihood of damaging the bird, which is more robust. If the goslings are accustomed to being handled it is a stress-free operation. They can also be close-ringed at the same time.

Sit down and put the bird across the knee. Rest it on its back and hold it against your body. The tail can be pushed downwards and the sphincter muscles gently rolled open, using the forefingers of both hands, to reveal the sex. Males have a small, 3mm long, curved, pale penis in paler-coloured breeds. It may be darker in Brown Chinese. Take care that this is not confused with the genital eminence in the female, which is a small, pale, raised blob on the outer edge of the vent.

When opening the vent of the bird to determine the sex, force should not be applied. It must be allowed to relax its sphincter muscles before applying gentle outwards pressure. Do not put pressure on the abdomen; these are small, delicate birds. If the sex cannot be determined after a couple of tries, then it should be left until another time so as not to cause bruising or undue stress. Some birds are easier to handle than others, and females generally have a softer vent which will open more readily. An experienced person should show you how to do this, rather than distress the bird.

If vent-sexing has not been practised at this early stage, it will be more difficult when the bird is juvenile, especially with larger breeds such as the Toulouse. Larger birds are more difficult to handle, and more distressed by the process, than a well handled gosling. It is also more difficult to see the tiny penis in the males at the juvenile stage, and it can be confused with the genital eminence in the female. Only when the males are sexually mature (twenty-six to twenty-eight weeks onwards) will the penis have grown to its adult size.

The legs of the birds are surprisingly large at four weeks, and they can be close-ringed or marked with spiral rings. With practice, vent-sexing gives complete accuracy. Confirmation can be by behaviour, size and voice as the birds grow.

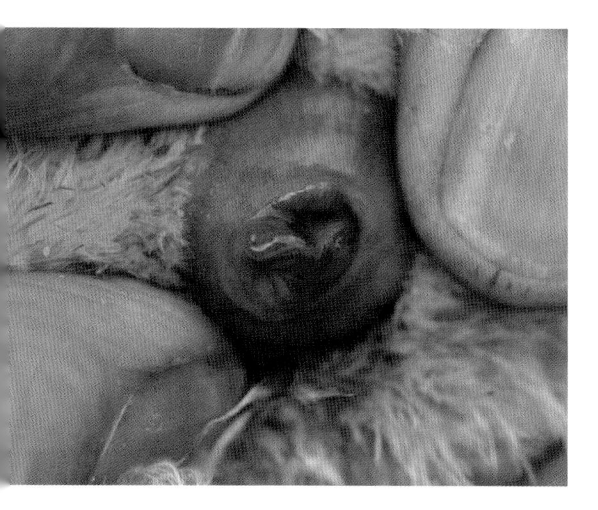

*The juvenile male penis is a tiny organ only about 3-4mm long in goslings up to sixteen weeks old and can easily be missed, especially if the organ is the same colour as the area around it.*

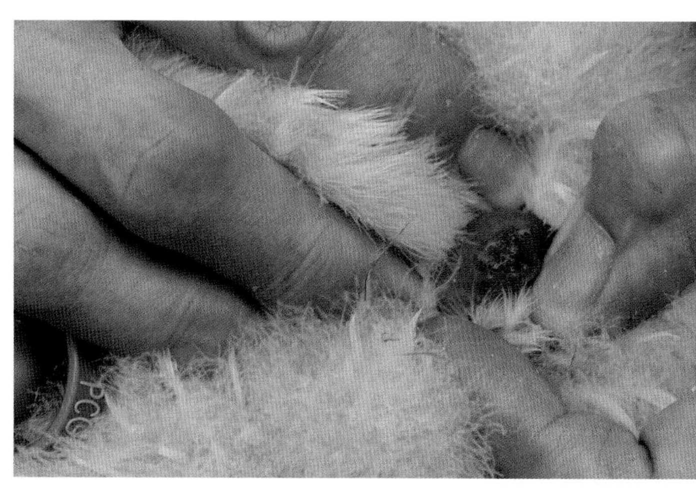

*The female vent lacks a penis and shows pink corrugation.*

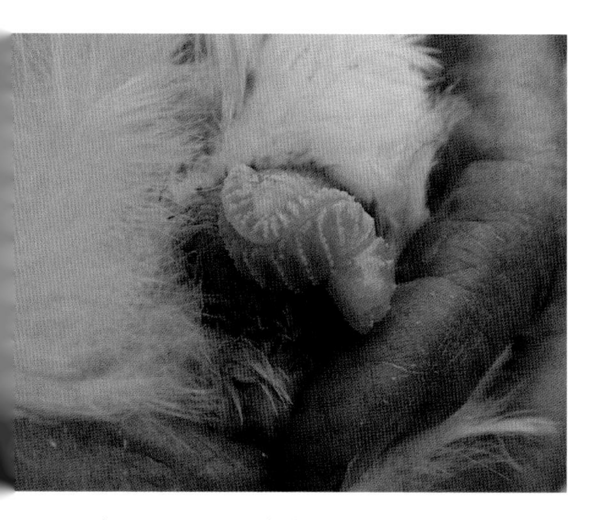

*Vent sexing: a mature gander is held upside down; the tail is bent backwards towards his spine and the vent exposes a pale, spiral penis, ¾-1in (2-4cm) long.*

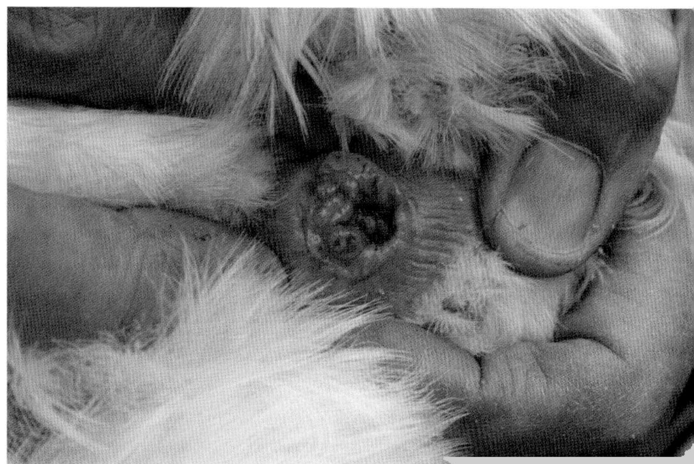

*Vent sexing: a mature female.*

## Identifying Birds

For a pedigree breeding programme, geese need to be marked: it is essential to know which birds are the parents, which are the offspring, and to mark any new stock birds brought in. Adult geese are often identified by using plastic spiral leg rings or flat plastic coiled bands. The latter are easily pulled off, however, and the spiral rings have been known to slip over the heel and dig in when the wrong size was fitted.

Closed rings are safer and better. Breeders put closed rings on their birds at around four weeks of age, when the closed plastic ring, numbered and dated, is slipped over the foot. It usually stays on if the timing is right. As the foot grows the ring will stay on permanently. These permanent markers are not removed for shows. Conventionally males are ringed on the right leg, and females on the left.

*A closed ring gives the date (2008), size H (1in/24mm) and has a unique number; the colour also differs from year to year.*

## CHECKLIST FOR BREEDING AND REARING EXHIBITION GEESE

It seems to be a common complaint that geese are not as good as they used to be. Maybe there is some validity in this view. In the earlier part of the century, geese were more commonly kept as farm birds or as exhibition birds by professional breeders. Only the latter birds were frequently seen at the shows. Now that geese and ducks have become more popular on smallholdings and as pets, which do get to the exhibitions, stock is bred and reared in a far greater variety of ways. The birds reared by the professional breeders with years of experience still make the size they should, but quite often stock in the medium- and heavyweight breeds is too small. Unless good practice in breeding and rearing is followed, then geese do get poorer by the generation. The following checklist is a summary of the recommendations for breeding quality birds.

* In-bred birds may produce smaller offspring; use unrelated stock.
* Parent birds need a nutritious diet of good quality grass, supplemented with breeder pellets. The heavier breeds such as Toulouse benefit more than lighter breeds.
* Do not use yearling females for breeding quality birds: the eggs are below the weight typical of the breed.
* Choose eggs of average weight with a good shell.
* Use clean, unwashed eggs. The goslings are less likely to get a yolk sac infection, and will grow well.
* Use eggs stored for no more than ten days, or less. Old eggs may hatch but can produce small birds.
* Eggs hatched early and mid-season generally produce larger birds than late season eggs. Use broodies to hatch these early eggs. However, note that early hatches (apart from the first couple of eggs) tend to produce more males; later season eggs are likely to have a higher proportion of females.
* Goslings need supplementary brewers' yeast plus clean, clipped greens if you cannot get them out on grass. This avoids leg problems.
* Graze the goslings on clean grass that has not been used recently by adult geese.
* As well as grass, provide ad lib growers' pellets/wheat (except whilst the blood quills are sprouting).
* Always have coarse sand and grit available to facilitate digestion in the gizzard.
* Worm the birds if they seem light - pick them up frequently.
* Heavy breeds of geese continue to grow until at least two years old, and must be fed on quality food during this period.

# 21 Commercial Geese – Table Birds

## Introduction

Goose has become a popular organic choice for the UK Christmas table market. However, the price per kilo is many times greater than the broiler chicken, slaughtered at only forty days of age. Geese for the premium Christmas market can be eight to nine months old, so the costs of production are obviously higher. This is a luxury food.

Goose production is a very specialized business: it is skilled, highly seasonal, and you do need a love of the goose. The best farmers always know their stock – it comes down to empathy and experience. So always start off in a small way with a pilot study, and take advice from a professional group such as British Goose Producers: they have the skills and the contacts to help their members.

Also, you do need an organization in a crisis. Geese are difficult stock to protect in the event of avian flu, largely because they simply cannot be temporarily isolated – that is, housed – from wild birds. The confinement of geese is not desirable on welfare grounds.

Avian influenza has not gone away from Egypt and the Far East, although the incidence of bird flu has died down with a better understanding of hygiene and disease vectors, such as the commercial transport of poultry industry products. Consideration of location is therefore important. Don't make the mistake of the turkey farm that set up adjacent to a wildfowl lake in East Anglia: it was insinuated that wild birds were the source of a 'flu outbreak, although they may not have been – nevertheless thousands of commercial birds were culled as a result of the outbreak first noticed in the turkeys. Thus there is no point in inviting a problem by locating a commercial operation close to an open body of water: although avian flu is carried by industrial vectors, wildfowl may occasionally carry it (there are many strains), and they are a source of duck viral enteritis.

Note that keepers of more than fifty birds in the UK must be registered with Defra, and one questionnaire item is about the distance of open water bodies from the holding.

Commercial goslings are sometimes imported from the continent. Occasionally, outbreaks of goose parvovirus (Derzsy's disease) are reported in association with these imports – for example as in 2004 and 2008. Vaccination is often used in goose-producing countries abroad, and has also been used in Great Britain following the GPV outbreak in 2004 (*see* Irvine *et al*, 2008). Although vaccinated goslings are protected, they can unfortunately remain carriers of the disease. To avoid disease transmission, Defra recommends that goslings from different flocks are not mixed.

Geese which have suffered from the disease can be sub-clinically infected, so any survivors of infection should never be used as breeders because they may transmit the virus through their eggs. Breeder geese are therefore best obtained from guaranteed disease-free UK stock. Commercial goslings are best bought as a batch and all used for the table – an all-in, all-out policy.

Commercial operations must also comply with Defra's five freedoms for farm livestock: freedom from hunger and thirst/from discomfort/from pain, injury or disease/from fear and distress/and freedom to express normal behaviour.

Specific guidance will also be needed regarding on-farm welfare/welfare at transport/welfare at market/welfare at slaughter. Find out more about these regulations designed to protect birds and animals at www.gov.uk/government/publications/poultry-on-farm-welfare

## Obtaining Stock

### Commercial Strains

In contrast to exhibition pure breeds, commercial birds are distinguished largely by their productivity, so characteristically they have a wide gene pool and have been bred for hatchability. They have generally been produced from a number of different lines of pure breeds in the first instance, but look fairly uniform by the third and further generations. Should productivity fall, a similar outcross with a different gene pool is used. These types are mass-produced, are not particularly distinctive, and are generally white. They are not usually included in standards books, which describe the characteristic appearance of distinctive types.

## Strains for the Table

In the UK, the **Danish Legarth** is a popular table bird. This is a strain of white goose which has been tailor-made to have a rapid growth rate with maximum feed conversion benefits. These geese attain 9.5–14lb (4.3–6.35kg), including the weight of the giblets. There are also larger commercial white strains reaching oven-ready weights of 16.5–17.5lb (7.5–8kg) at twenty-four weeks.

A table strain which has been recommended in the USA is a **White Chinese crossed with an average-size white Embden gander** (*see* Acheson, 1954). The egg-laying capacity and hatchability comes from the Chinese, and the slightly larger size from the Embden.

**White Chinese** themselves make a good meal: their flavour is preferable to that of the Brown, and the carcass looks better. However, the ganders need to be killed before maturity (preferably before sixteen weeks) otherwise they develop a stronger flavour. Browns are particularly difficult to pluck and dark fluff and pins are left behind.

**Roman** geese were first introduced into Britain as small table birds. They were good layers, and their plump, round bodies were ideal for small ovens.

The **Brecon Buff** is a hardy, economical farm goose for small-scale production. Larger than the Roman, it carries a relatively large amount of muscle for its size compared with other breeds, and can be fattened up easily after grass rearing. The soft buff feathers pluck out more easily than white feathers without tearing the skin, and leave a clean-looking carcass. It dresses out at 7–12lb (3–5.5kg) in weight, depending on the strain and feeding. Of course true Brecons are a rare breed and are certainly not table birds now. However, a **Brecon/American Buff cross** would give all the advantages of buff plumage, and should also produce strong goslings.

Pure-bred Embdens and Toulouse are unsuitable as commercial geese because pure breeds are more difficult to hatch, and these large birds do need special care and more feeding. Also, in the Toulouse, the gullet and keel is wasted weight, whilst the Embden has a large frame with plenty of bone. Commercial Toulouse geese in France are very different; they are a lighter-weight version of the English exhibition birds, and have a wider gene pool.

Commercial heavyweights are often an **Embden–Toulouse cross**; this produces an attractive, light-grey bird whose feathers shade to white on the neck. These are fast growers and outgrow their parents. Do not be tempted to use these first crosses (that is, inbreeding using siblings) for breeding stock for the following year's goslings, because it is the pure-bred parents that produced the first cross vigour. If, however, the parents were poor layers, it would be worthwhile seeing if the female offspring were better, and using them in a subsequent programme. It was the Toulouse–Embden crosses in Victorian Britain that increased the egg-laying capacity of our English Embdens to thirty-five instead of the German strain's total of fifteen (*see* Parr, 1996).

The disadvantage of opting for a larger table breed is that a higher ratio of ganders is needed for fertility, and the ganders also need to be replaced more often than with light breeds. Older, overweight birds become infertile; five to nine years is the recommended breeding span for commercial ganders. If several geese are to be kept in a group for flock mating (as opposed to breeding pens), a ratio of 1:5 is possible with young light breeds – but even in flocks, the birds will tend to settle eventually in pairs and trios as they get older.

## The Availability of Goslings

The traditional saying is that a goose should lay by St Valentine's Day, when light levels reach ten hours daily (*see* Chapter 18). So a few birds are hatched as early as late March, and these first birds are ready for the Michaelmas market (29 September) at around twenty to twenty-four weeks of age.

Quantity production of goslings will only come from breeds which are prolific layers, so those producers who wish to rear a flock of a hundred probably do not keep their own stock geese, but buy their goslings at day-old or week-old from commercial producers. In the UK these are mostly located in eastern England, and can be found through the BGP.

This separation of hatching farms from rearing farms is sound practice. The advantage of not having parent stock on the ground over much of the winter and early spring is that potential problems from gizzard worm and coccidiosis will be minimized. The ground is rested from birds from mid-December until even the beginning of May, when the first goslings are let out from the rearing area. This really does help to reduce the incidence of parasites and disease. In addition, the day-old goslings arrive parasite free, and can maintain that condition, with minimal worming on Flubenvet, when the ground they graze has been depopulated for several months each year.

Hatching geese is still a very specialized – and personal – skill. Hatching is probably easier in the eastern counties where rainfall and relative humidity is lower than over most of Britain; most commercial goose producers are still located here, or in the English

Midlands. This is a long-established specialization, perhaps going back to the domestication of the goose from the Fenlands, and the tradition of walking geese to the London market from the main production areas in Norfolk and Suffolk (*see* Chapter 1).

## The Management of Commercial Geese

### Caring for Young Goslings

Goslings are normally obtained at 'day old'. There are sound reasons for this, the main one being that they have sufficient food in their internal yolk sac for the next twenty-four to forty-eight hours. They are also small, can be transported easily, and can keep each other warm.

Where large groups of up to 200 goslings are reared – a larger unit is not recommended, and smaller groups are preferable – heat facilities must be good. Crowding under heat lamps must be avoided, and a large draught-free area with industrial-scale heaters – for example of the gas brooder type supplied by Maywick – is essential.

The birds are normally reared for two to three weeks in a vermin-proof, closed, heated space with feed and water available at all times. The heat is reduced over this period. Additive-free starter crumbs for waterfowl are fed for the first two weeks, and from then on, pellets are gradually added to them. Water must be freely supplied in ball-valve drinkers suited to the size of the birds so that they cannot drown. Problems with 'grazing' each other may arise if green food is not made available. Also watch for vitamin deficiency if the birds are kept indoors: supplementary brewers' yeast is beneficial (55mg nicotinic acid per kilogram of food is recommended). Note that continued high protein levels are not good for geese: lack of exercise, too much protein and lack of green food will result in leg problems in intensively reared birds.

Birds go out to clean, rested pasture when they are large enough to be safe, and the weather is suitable. A useful arrangement is to have a series of grass pens that can be used in turn. Grass becomes a major part of the goslings' diet, though at night and in wet weather they should be returned to their own well aired polytunnel or shed: they need to be kept dry overnight until they are feathered on the back.

Overnight, pellets are generally available ad lib, with a constant water supply. Wheat is gradually introduced and increased to around 50 per cent at ten weeks

*Geese move to protected areas at night.*

*A constant overnight water supply.*

*Night-time enclosure with feeding troughs.*

of age; grit and coarse sand must always be freely available.

If the grazing is good, geese will voluntarily reduce their compound food intake, and at about four to eight weeks of age can get by on just under 3oz (80g) per day, rising to around 5oz (140g) at nine to eighteen weeks. The actual amount consumed will vary according to temperatures, grass quality, disease control and breed.

Well feathered geese can be kept outdoors overnight as long as they have a fox/badger-proof compound, and shade from early morning sun. A lot of litter, such as chopped straw, is often used in the compounds, and this goes back on the land in rotation with cattle. Although swimming water is often not available, good litter and extensive free range on grass keeps birds clean.

Goslings should be reared and kept in their hatching groups throughout their life: this routine keeps them happy and stress free, particularly if they are familiar with their handler. Upsetting a group will adversely affect the result, particularly in the fattening period, when removing even just a few of the birds can upset the group dynamics and put birds off their food.

## Feed Conversion in the First Ten Weeks

Goslings grow most rapidly up to ten weeks of age. By that time, an 8.8lb (4kg) bird may have eaten 26lb (12kg) of food, if fed ad lib, giving a conversion rate of bird:food of 1:3. Note that all strains of geese do not grow at the same rate: Chinese are smaller and will therefore consume less food than strains that will reach twice their weight.

As the bird gets older, the growth rate reduces but the bird still needs food for its own metabolism, thus the food conversion rate becomes much worse. Edward Brown observed, in 1913:

> Apart from green goslings, I question whether geese can be grown profitably unless they find in a natural manner practically all their food from about eight weeks onwards to the time when fattening commences. In this may be included spilled grain when placed on the stubbles.

This is why intensively fed food-chain animals are generally slaughtered at the point where they have maximized their growth for minimal food input. Whilst the cost of basic feedstuffs such as wheat continues to rise, it is now very important to use strains of animals and birds which are good converters of basic foods, and to use on-farm products.

## The Middle Stages of Production

Intensive feeding throughout life is both undesirable and uneconomic for Christmas table geese. The goose is a very good converter of high-fibre feeds, unlike poultry which must be largely grain fed. That is why, for centuries, they have been the ideal smallholder bird.

Christmas geese are up to eight months old when they are ready for slaughter. Figures quoted in Stevens (1989) show that birds fed intensively during the middle period of production were heavier. However, they carried much more fat in the body cavity at slaughter than birds finished for just the last few weeks. Muscle weight was similar.

To keep costs down mid-production, a large part of the diet should come from home-grown fodder. Feed can be any succulent grass with some clover, but also a variety of vegetable crops can be used, such as stock-feed carrots, potatoes, cauliflower trimmings. In warmer climates alfalfa is sometimes grown, but is not as palatable as grass.

The birds should have extensive range on pasture well grazed by cattle and sheep, in rotation. Geese cannot be expected to survive on restricted, dirty grazing: they will suffer from coccidiosis and leg infections, and there will be a build-up of parasites. Breeders who stock intensively avoid these problems by feeding the birds almost all of their diet from pellets and wheat, which is part of the reason why pure breeds from breeders can be more expensive than table geese.

The stocking density for geese on pasture will vary depending on the soil conditions (sand or clay), the quality of the pasture, the amount of rainfall, the time of year and rate of grass growth, the amount of supplementary food, and the age and size of the geese. Figures vary widely between fifty to 125 birds per hectare for growing geese, and twenty birds per hectare for breeding geese (when the rate of grass growth is slower in early spring). The lower figure is preferable for growers because the grass stays cleaner, and this will avoid coccidiosis.

### Finishing

Swimming water can be allowed during the mid-rearing period but not during the finishing period. This is because more calories are consumed by the birds from frequent contact with water and exercise.

Between four and seven weeks should be allowed for finishing, depending on the condition of the stock. In larger operations, range is often restricted (to reduce exercise) to an enclosed yard, but in smaller farm operations the geese are often allowed to graze in a reduced grassed area whilst grain and fattener (unmedicated) pellets are also fed ad lib. The area must be kept clean; a series of small grassed pens or a well littered yard are needed to avoid disease. The geese must be kept clean and feel secure.

Wheat is generally used as the cereal grain; maize makes the birds too fat. Although barley is sometimes used – the grains are rather spiky and it is best bruised. Ad lib grit and sand must also be available.

The flavour as well as the weight of geese is also improved by feeding cereals. Geese fed exclusively on grass may be plump enough but can taste 'fishy' – although grass-fed birds are probably nutritionally better, consisting of more 'healthy' fats than grain-fed birds.

Some geese are marketed earlier than Christmas as plump, mainly grass-fed, tender summer geese. The traditional time was at Michaelmas when the geese had gleaned the stubbles – making sure that grain was not wasted – and had put on more weight.

## Slaughter: Regulations and Procedures

If the birds are for personal use, you are not bound by regulations relating to carcass preparation, but you are obliged to slaughter the birds in an accepted, legal way. Old-fashioned methods of bleeding the birds to death are illegal.

Ideally the birds should be stunned by an electrical stunner causing instantaneous insensibility, and then bled or the neck dislocated. Larger birds such as geese can be up-ended prior to stunning in an open-bottomed cone to control their movement, especially the flapping of their large, strong wings. It is advisable to obtain training before using this type of equipment for any purpose, whether commercial or private, and the Humane Slaughter Association (HSA) provides one-day training courses.

Other legal methods of slaughter currently (2011) include decapitation or neck dislocation, although the HSA has reservations about both methods:

> In the case of neck dislocation, it is difficult to consistently achieve an immediate loss of consciousness. Similarly, decapitated birds may continue to show brain activity for up to 30 seconds after the cut is applied. As such, there is potential for distress, pain and suffering. The HSA recommends that neck dislocation should only be used in emergencies or for very small numbers of birds where no better method is available. You do not require a slaughter licence to carry out neck dislocation or decapitation on the premises on which the birds were reared.

### Procedure

The birds should not be fed overnight prior to killing to allow up to twelve hours for the gut to be cleared of food. They should be allowed water. This makes preparation easier and also more hygienic, as the gut is empty. Then small groups of birds should be separated from their normal flock and environment. Quiet handling in small groups, by their normal keeper, saves panic and distress, and is also a better working environment.

For a few table birds, neck dislocation can be practised. Dislocation is done by hand with smaller ducks, but the length of the neck and body of the goose makes this impractical, so the following method can be used:

* Hold the bird by the legs, its breast away from you, and its head and neck laid out across the ground facing towards you. The crown of the head is uppermost. Place a pole, such as a new broom stale, across the back of the neck.
* First, applying only restraining pressure, elongate the bird's neck, legs and body until they are tight whilst you are standing with bent legs and straight arms.
* Only when you are ready, stand with your full weight on the pole with one foot on each side of the bird's neck. The dislocation will occur by straightening the legs; thigh muscles are much stronger than arm or back muscles.

The distance you need to pull is very small. The vertebrae are dislocated quickly and easily and you need not fear pulling the head off. Pulling at 90 degrees to the line of the head clicks open the vertebrae and severs the spinal cord. Sometimes the blood vessels will rupture so that blood spills out of the bird's mouth, but the bird is already dead at this stage.

I have also seen this method carried out with the goose presented the other way around – that is, with the bird's breast facing towards you and the head facing away.

I should emphasize that if you are unfamiliar with this method you should be shown how to do it by an expert. It is not the kind of thing that you would want to practise unsuccessfully.

FURTHER INFORMATION

For more information and for courses about slaughtering methods, go to the HSA website www.hsa.org.uk. Further details about courses and slaughter licences can be obtained from the local Animal Health Office: *see* Defra's Food and Farming section at www.gov.uk/government/publications/poultry-on-farm-welfare for up-to-date information.

If the dressed birds are to be sold, also check the current regulations for the facilities required in the preparation area, and in marketing, as there have been frequent changes in legislation over the past few years: *see* www.food.gov.uk (Food Standards Agency) regarding Meat Hygiene Legislation.

The HSA notes:

> Dislocating the neck of the bird may cause rupturing of the spine and concussion. When done correctly, this results in the bird losing consciousness immediately and irrecoverably. However, it is difficult to achieve concussion consistently using neck dislocation, therefore this method is not suitable for routine commercial harvest. It may, however, be appropriate for small numbers of birds or for emergency killing. . . It is important that a technician applying neck dislocation is mentally prepared to carry through the whole procedure. Practice on dead birds may improve the confidence in application of this method. The various means of achieving neck dislocation are detailed in the HSA publication, Practical Slaughter of Poultry 'A Guide for Small Producers 2nd Edition' (pages 17–20).

## Plucking and Evisceration

Plucking and dressing the birds is no light task. Dry plucking is needed for high quality goose down, but this means plucking the bird twice because the large outer feathers are removed first, then the down. In commercial operations, this is generally done while the carcass is still warm, and the plumage is removed in one go, but the skin is more likely to tear than if the bird is left to cool. If quality down is required with no soft, fleshy pin feathers to spoil it, the birds must be in full feather and at least sixteen to eighteen weeks old. Alternatively they can be killed when they have completed their first set of feathers. Appleyard (1933) recommended killing Roman geese when just in their first feathers (Stevenson, 1989, cites nine weeks) for economical table production. Food conversion rates are then at their most favourable.

Avoid killing birds when they are moulting their first set of juvenile feathers at about twelve to sixteen weeks old. Assess the stage of feather development by parting the feathers and checking for any quills filled with pulp first.

Electrically driven plucking machines can be used with geese, but many small-scale producers use wet-plucking methods. A small, open-topped boiler is used to heat the water to about 65°C (149°F). Then the bird is hung by a piece of string tied to the legs and dipped in the water for about 1½–2min. Note that older, mature geese may need more time, or slightly hotter water. To ensure that the water penetrates the feathers, ruffle them up under water with a stick. Try this, then test the feathers to see if the time is correct or if re-dipping is needed. Feathers and the down should easily pluck or roll off, leaving a clean carcass.

The advantage of using water at this temperature rather than boiling is that it is safer for the operator, and there is more leeway over the timing. The ADAS (Agricultural Development Advisory Service, www.adas.co.uk) 1981 leaflet, for example, recommended even up to three minutes.

Where wet feather and down is produced, it must be washed and dried before use, and flies excluded (*see* Acheson, 1954, for methods of handling).

For larger operations, mechanical dry-plucking machines are used, and are much quicker. The dry-plucked carcass also keeps better whilst hung long-leg (not eviscerated), for a longer period of time, in an airy place at 0–2°C. Chilling the carcasses rapidly prevents bacteria growing, and hanging will improve the flavour and tenderness. Machines are also used to pull out the leg tendons of the geese, so that the meat is easier to serve at the table.

For the commercial finish, the wax-dip bath is essential, where the plucked carcass is dipped in liquid, warmed poultry wax at 65–70°C. The carcass is then lifted out of the bath and cooled so that the wax hardens: it is then stripped off, removing any remaining pin feathers and down which are almost impossible to otherwise remove by hand. The wax is then re-used by re-melting, and straining out the feathers, and sterilizing, if necessary.

Smaller-scale operations use scorching instead, but the wax (of the correct type) gives a neater finish. Singeing of the feathers (mostly down and wispy filoplumes) can be done with a gas blowtorch, but can give a black look and burned flavour.

After hanging, evisceration should finally take place in a cool, clean area. For personal use we put the carcass on to newspapers so that the eviscerated contents can be rolled up and leave no mess. Correct disposal of waste is essential, and commercial packaging is very important to preserve the carcass and for good presentation of the product. Note that the loss in weight of the carcass is approximately one third, excluding giblets.

## Commercial Constraints and Marketing

Commercial suppliers like to market some of their geese for Michaelmas and over the autumn period to advertise their product, and also to reduce the Christmas peak: the once-a-year processing rush is a great strain on the seasonal market supplier. By necessity, processing depends on seasonal labour, so contacts must be good to operate a large business. It is unlikely that local labour will be available, and large organizations are often dependent on migrant labour, with its attendant demand on accommodation. Running a livestock business largely dependent on the Christmas market is a huge risk. There is also the capital investment in facilities and buildings that are used mainly for only three weeks of the year in early December.

Processing and marketing of the Christmas goose and its products is a highly specialized task. Good marketing is an essential part of any business, and in this one, with a highly seasonal and perishable product, marketing is the key. Packaging and presentation are very important, plus the advertising of recipes: a great deal of hard work goes into advertising the product at the autumn food fairs, exposure on television slots, and at specialist food outlets. Marketing the goose also includes the by-products such as goose fat, feathers and down. Goose fat is a luxury product for roast potatoes at Christmas, and some goose producers also market their own ethically produced feather and down products (*see* Chapter 3). The dry-plucked feather and down is processed, returned to the pillow makers, and sold at farm shops, guaranteed UK-produced.

Producers market their down through The Norfolk Feather Co. or Shingfield & Sons. Both companies are in the traditional goose country of eastern England.

## Geese for the Table

Over-fattened geese, force fed in confinement, are obviously fattier than birds that have been out on range. Grass-fed birds that have been able to exercise are leaner: the folds of the intestines are separated by less fat, and the gut itself is different in quality, being much greater in diameter than that of the corn-fed bird. However, the older grass-fed birds tend to be a bit tougher and less tasty. You will have to decide for yourself on the type of product for your market. The quality of the feed will also determine whether it can be marketed as 'organic' or not.

Goose meat was originally alleged to be full of cholesterol, but opinion seems to have changed. It has been observed that the rural French eat duck and goose quite frequently and that they seem to suffer no problems.

Because of its high fat content, goose meat is very rich and filling. Servings can therefore be quite small, and this is fortunate because goose is so expensive per kilogram, given that a good deal of the weight is bone. A small goose weighing 10lb (4.5kg) should make a good Christmas meal for six – allow approximately 1lb–1.5lb (1kg) per person – but the amount of meat will depend upon the breed and how well the bird has been reared. For example, a small white Chinese goose should provide enough meat for six, as long as it was well fed.

## KEY POINTS

* UK geese are mainly reared for the Christmas market.
* The trade is highly seasonal and specialized.
* Guidance with animal welfare regulations should be sought.
* Commercial strains must be used for hatchability and good food conversion rates.
* Goslings are often bought from specialized hatcheries.
* Make sure that stock is disease-free; enquire if the birds are GPV vaccinated.
* Do not mix commercial geese with pure breeds, to ensure biosecurity for pure breeds (see GPV, Chapter 22).
* The laying cycle is mid February to late June.
* Goslings are mainly hatched from March to July.
* Specialist facilities are needed for rearing and protecting the goslings in the first few weeks.
* Goslings in first feathers can be slaughtered as broiler geese at around nine weeks.
* Michaelmas and Christmas geese can be twenty to thirty-four weeks old before slaughter.
* Geese must therefore be fed on cheaply produced maintenance rations in the middle period of production when they are in store condition.
* Fattening is essential for the quality Christmas product.
* Training for a slaughter licence is recommended.
* Note UK legislation for 1 January 2013: manual dislocation will only be permitted on birds of less than 3kg, up to a maximum of seventy birds per day. Most geese are over this weight.
* Contact the Humane Slaughter Association regarding satisfactory stunning/despatch methods.
* Food Standards Agency regulations apply in commercial food preparation areas.
* Good marketing is essential for this seasonal, perishable product.

# Part VII
# Keeping Geese Healthy

## Introduction

Disease prevention is far more important than attempts at disease treatment and control. Geese are generally healthy birds and do not suffer from many ailments.

Be observant: good stockmanship is an art which is acquired through interest in the birds. Watch stock closely for signs of unusual behaviour. Birds are very good at concealing the fact that they are ill, until it is too late. If any bird is behaving differently from usual, pick it up to have a closer look at its eyes, nostrils and vent, and to check its weight. When birds have to be collected and driven into a shed each night, rather than being left out in a fox-proof pen, symptoms of illness are more likely to be recognized.

Advice will be needed on choosing correct medication. Coccidiosis is not caused by bacteria, for example, and therefore does not respond to antibiotic treatment. Diseases caused by bacteria do not always respond to every antibiotic; the correct one is needed for specific conditions. Many treatments are 'prescription only medicines' (POM), purchased from and used with the advice of a vet. Note that some of the preparations used for treatment have not been specifically developed or tested for waterfowl, and for some coccidiostats the dosages may be for other avian species. Similarly, some wormers have not been specifically tested for geese. Dosages are based on practical experience, but adverse effects may occur and should always be watched for.

It is often impossible to tell why a bird has died unless a vet does a post mortem and then, if necessary, sends off a tissue sample to a pathology laboratory. This is a time-consuming and expensive procedure, and unless there are several birds dying inexplicably, not worth doing – and by the time the result is received it is often either too late, or the problem has resolved anyway. In contrast, a local vet can usually do a quick check for TB, gizzard worm or the likelihood of coccidiosis. The advantage of having a laboratory report, however, is that the correct treatment can be prescribed if the symptoms recur in other birds. Disease must be caught early if treatment is to be successful.

# 22 Ailments and Diseases

## Introduction

Birds that are ill should be handled, and checked for weight, parasites, abnormalities and type of droppings. Those that are thin or coughing should be wormed as a matter of course. Unwell birds frequently appreciate coarse sand, and should always have access to it. Following a few simple guidelines, as in this checklist, will reduce disease to a minimum:

* Noticing signs of illness early saves birds. Medication is more effective. The sick bird can be isolated (though in the case of geese, with a companion), reducing chances of further infections.
* Provide clean water. This allows birds to wash their eyes, to clean their feathers and preen effectively.
* Make sure that all food is stored correctly, in date and free from moulds. Do not leave food in dishes to go mouldy. If possible, buy pellets manufactured for waterfowl (for example ducks), not hens.
* Provide sand and grit for the gizzard.
* Keep the night quarters dry, bedded with white wood shavings, and well ventilated. Remove all foreign objects which could cause injury, such as string, plastic, fragments of wire. Check for hooks and nylon line in fishing areas.
* Check birds for external parasites - northern mite and even leeches and ticks. Worm birds if they are thin; otherwise worm routinely once or twice a year with flubendezole in Flubenvet, which treats all types of worms. Alternatively, use a single-dose drench from the vet.
* Make sure the birds do not come into contact with poisonous materials, such as rat poison, creosote, lead shot, weed killer, antifreeze, nitrate fertilizer (in granules or as run-off from recently treated fields), poisonous plants.
* Do not allow access to rotting animal and vegetable waste, septic-tank waste or farm slurry.
* Provide shade and plenty of clean water in a heatwave.
* Protect from frostbite in severe winter weather.
* Handle birds quietly; do not feed immediately before catching.
* Rear young birds on clean ground which has been limed and rested. Rear them separately from adults to avoid high concentrations of pathogens before young birds have had time to acquire immunity.
* Remove all manure from the premises if possible; spread on land used by other types of animals.
* In commercial production especially, where large numbers are involved, be scrupulous in cleaning sheds, yards, feeding and drinking equipment.

## A–Z of Ailments and Diseases

### Air Intake Beneath the Skin: see Subcutaneous Emphysema

### Aspergillosis

**Cause:** Spores of various fungi proliferate in damp, mouldy bedding, particularly hay and straw. Moulds can also develop on damp food, particularly pellets and waste grain.

**Symptoms:** The fungus grows inside the respiratory system and gives the symptoms of pneumonia: rapid breathing, gasping for breath and drowsiness, loss of voice. The spores also cause respiratory distress (farmer's lung) in humans.

**Treatment:** A cure is unlikely. Fungicides such as Nystatin have been used in high-value birds, but the problem should be avoided by good management. Remove environmental challenge; keep birds in clean surroundings. Use whitewood shavings for bedding. Clean straw and shavings should be used in the breeding season for nests in a shed, and bark peelings for damper places outdoors. Nesting material must be clean, otherwise aspergillae can be transferred to the incubator via the eggs and infect even the embryos and the goslings. Store food in dry conditions and never feed mouldy bread, wheat or pellets. Keep birds off land where mouldy waste agricultural products have been spread

### Avian Tuberculosis

**Cause:** Tuberculosis (TB) can be carried by wild birds. It is rare: only one case has been identified in thirty years in a bird bought in, which transferred the disease to only one other. This was fortunate, because this condition can be very persistent.

**Symptoms:** The disease takes a long time to develop, and the bird will become listless and thin, and will die. A post mortem will reveal infected joints and an enlarged liver, studded with small white or yellow nodules.

**Treatment:** All associated birds should be tested (very much as a human TB test). The pin-prick test to the skin should be applied to the side of the bill at the fleshy base where it is attached to the head. Do not allow the feet to be used as the bird can go lame. Affected birds react by a swelling at the test point and must be culled and burned. Further testing is necessary at one- to two-month intervals until the flock is clear. The birds should preferably be moved on to clean ground, and the housing disinfected.

## Botulism

Otherwise known as 'limberneck'.

**Cause:** Toxins produced by the bacterium *Clostridium botulinum* in rotting dead animals, and rotting vegetation in boggy areas, especially in warm spells in summer.

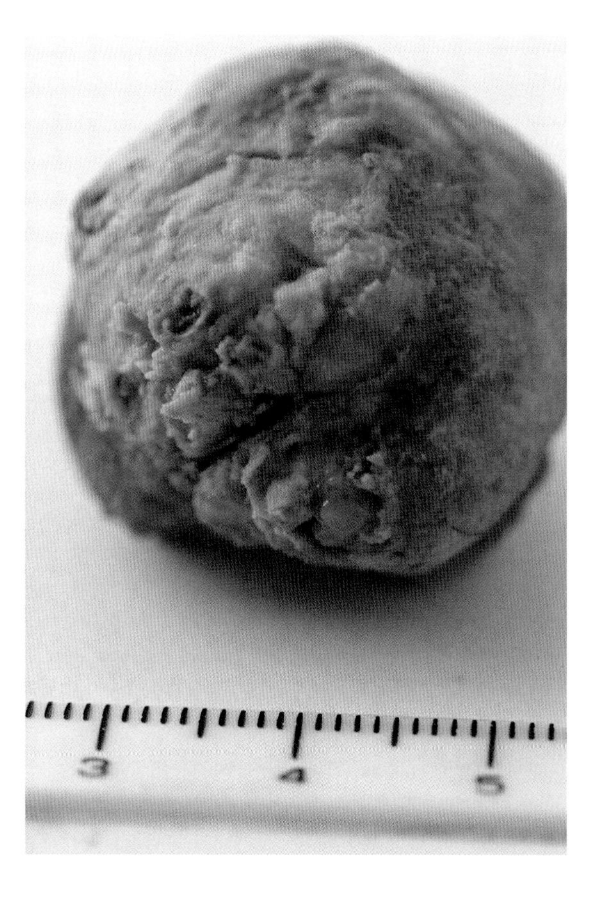


**Symptoms:** Loss of muscular control of the legs, wings and neck, hence the term 'limberneck'. Birds are unable to swallow.

**Treatment:** Exclude birds from rotting materials. There is no treatment for the disease other than removing the source of the infection and providing clean drinking water to flush out the system. Drenching is necessary if the bird cannot drink; Epsom salts may help recovery; antibiotics help stop secondary infection.

## Bumblefoot

This condition is identified by a hard swelling on the foot.

**Cause:** Bacteria, such as for example staphylococci, enter through an injury.

**Symptoms:** The bird limps as the swelling grows on the heel or under the toes. This can be pus and/or granular tissue.

**Treatment:** Treat a limping bird with antibiotic first. Remove any crust and infected material. Small lesions are encouraged to heal by access to swimming water, which also takes pressure off the foot. If the environment is dry, clean the foot and use antibiotic spray (POM) on the lesion.

Where the accumulation of granular tissue is advanced, antibiotics fail to work because the blood supply is cut off from the infected material. Surgical removal is best done at an early stage to excise the material cleanly, and before infection has entered the bones.

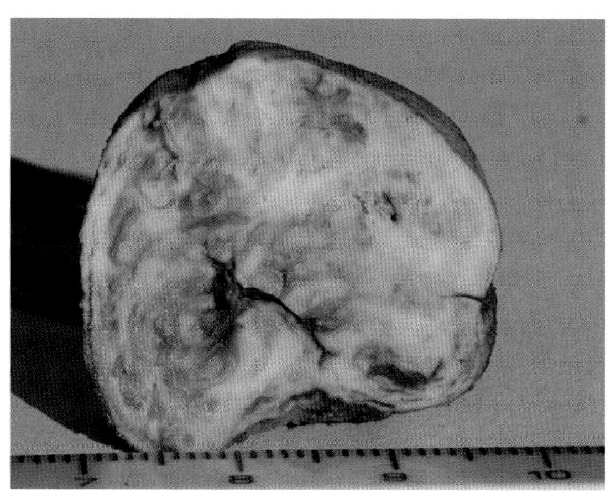

*A ¾in (2cm) ball of granular material removed by surgery.*

## Coccidiosis

**Cause:** Coccidia are one-celled protozoan parasites which are species specific. Geese can get two forms of coccidiosis: intestinal coccidiosis caused mainly by *Eimeria anseris*, and renal coccidiosis caused by *Eimeria truncata*, found in the kidneys. *Eimeria anseris* lives in the gut; it causes tissue damage to the intestine which interferes with nutrient absorption. Young birds fail to grow, become dehydrated and anaemic, and susceptible to further secondary infection and death. Geese are less affected by the disease than poultry, but it is a problem if they are grazed on dirty grass covered in droppings. It particularly affects young geese in the summer heat in damp weather when excreted oocysts 'sporulate' - become infective in warm, moist conditions. The life cycle within the bird is four to seven days. Active, 'ripe' oocysts picked up from the grass are swallowed and embed themselves into the lining of the intestine and replicate asexually for several generations. Eventually, new oocysts are formed through sexual reproduction, and these are shed on the grass to renew the cycle.

**Symptoms:** Intestinal coccidiosis causes lethargy, loss of weight and balance, and blood in the droppings which cannot be seen easily if the birds are eating grass. Look at droppings passed in the morning, just before the birds are released from the shed; any blood can then be seen on clean shavings or newspaper. Renal coccidiosis also affects young geese and can cause high mortality. Again, birds are depressed and weak, with diarrhoea and whitish droppings.

**Treatment:** A vet can gauge the level of infection from a droppings sample and the birds can be treated with an anti-coccidial drug in a single drench, or in the drinking water. The coccidiostat added to some poultry grower foods is not a treatment.

Oocysts are present on all poultry farms, and even the healthiest birds are likely to be affected with a few parasites. These birds are resistant to further attack. Fatal disease follows in birds which have failed to acquire resistance which normally follows a low level of infection. Young geese are much more likely to be affected than adults. Oocysts take two days to 'ripen' after they have passed through the gut, therefore using clean ground, and feed from bags, helps to keep down infection.

## Coordination Loss

**Cause:** Loss of balance and lack of coordination is rare in geese, and the cause is often unknown. It may be caused by listeria found in soil and faeces, during hot spells in summer.

**Symptoms:** Geese will often sit down when they are ill. They may lose their balance if they try to move around (*see also* Botulism).

**Treatment:** Prompt action with the correct antibiotic can be effective in some cases. In a hot spell one summer, young Chinese geese were eating soil and presumably picked up the disease from the ground. They lost coordination, ran a high temperature, and although their condition improved slightly with antibiotic, they eventually had to be put down. They continued to eat and drink throughout the illness. Pathological laboratory analysis confirmed listeria.

Birds will also go 'off their legs' if transported badly, if they have a terminal liver complaint, or internal tumours. Loss of balance in cold weather may possibly be caused by a viral infection of the spinal cord. Diagnosis is difficult, and in each case the bird should be thoroughly checked for weight and any abnormalities, including in the ears and at the vent.

## Crop Binding and Sour Crop

**Cause:** Geese do not have a crop to store food, unlike the chicken. Food is stored in the gullet and proventriculus in a similar way to the crop, hence the term 'crop binding' or crop impaction when food becomes blocked. At this point the bird is receiving little or no nutrition, and will starve. It should either be put down or receive immediate treatment.

**Symptoms:** Toulouse geese are more prone to this condition than other breeds, probably because of the shape of the keel which causes the proventriculus to be carried low down. A hard lump develops at the base of the keel if the food is held up in its passage through the digestive system. This lump cannot always be seen. If a bird fails to eat, is head shaking, and having difficulty in swallowing water, examine it to see if this part of the digestive tract is clear.

In addition to physical blockage, the warm, moist environment can cause bacteria and yeasts to multiply. The yeast *Candida albicans* is found in the environment and in normal, healthy birds, and can cause 'sour crop', or thrush, if food is held up. In such a case, the bird will have a sour smell to its breath, in addition to the blockage.

**Treatment:** If the condition is not advanced, first try lubrication with medicinal grade liquid paraffin: gently massage the lump to break it up, whilst holding the bird's neck out and downwards. The bird will have difficulty in breathing in this position, so do not persist. It must not inhale infected particles. Unlike mammals, birds cannot vomit, but they can eject waste from the oesophagus by neck shaking.

If this procedure is unsuccessful, the bird can be operated upon by the vet to clean out the impacted debris. Treatment with a fungicide and/or antibiotic may also be necessary.

Avoid problems by making sure that birds have access to coarse builder's sand for grinding grass in the gizzard. Also, keep the grass short so that it is tender; long, stringy grass will cause problems. Note that birds should not be allowed to gorge on sand and grit if it has previously been unavailable; it should be limited in quantity until they are used to it, because excess sand can also cause a blockage.

After treatment, ensure that food is regularly available in limited quantities, and that water is freely available. Cider vinegar in the drinking water is recommended as an antidote to the fungal infection.

## Derzsy's Disease – Goose Parvovirus (GPV)

**Cause:** Goose parvovirus is transmitted by subclinically infected adult geese. Such birds can act as carriers, excreting virus in their droppings. They also transmit the virus through their eggs, resulting in the transmission of infection to goslings at hatching. Gosling mortality can be as high as 100 per cent in birds under one week of age.

**Symptoms:** Goslings suffer from swollen eyelids, a necrotic lining to the surface of the mouth and tongue, loss of down and reddening of the skin. They drink profusely, lose their appetite, and have white diarrhoea. GPV is very contagious, causing high mortality in birds under five weeks of age. Survivors are stunted.

**Treatment:** There is no effective treatment for the disease. Avoid commercial goslings, and eggs, imported into the UK from Europe. Defra advises the following:

> Control relies upon good biosecurity, hygiene and eliminating carrier birds. If a GPV outbreak occurs, it is not advisable to retain recovered or in-contact goslings for breeding, as the birds are potential carriers of the virus and may transmit GPV via their eggs. Suspect carriers should be reared away from other geese and Muscovy ducks. On breeding units, only eggs from known GPV-free flocks should be incubated together and good hatchery hygiene should be maintained.

Vaccination is used in commercial birds in Europe. Goslings are vaccinated at two weeks, where the parent stock has also been previously vaccinated and the goslings have some immunity.

## Dropped Tongue

**Cause:** The lower mandible becomes full of food waste and grit. The condition can develop with a poor feeding regime – for example if birds are fed a dry mash with insufficient water to clear their mouth. It is also alleged to develop if grass is too fibrous.

**Symptoms:** The impacted waste causes the floor of the mouth to sag and the tongue drops into the hollow, causing problems in eating and drinking. The bird often shakes its head and neck; it also loses weight. This most commonly occurs in Toulouse, but it can occur in other breeds.

**Treatment:** Treated early enough (before perforation of the skin by abrasive material or infection) this condition can be cured by stitching the loose skin together (not the inner membrane of the mouth), then drawing the stitches tightly together. The skin becomes necrotic and will eventually drop off. This does work, but infection is possible. Stitching, surgical removal and cauterization by the vet is preferable.

## Duck Virus Enteritis (DVE)

Otherwise known as 'duck plague'.

**Cause:** Ducks, geese and swans are susceptible to this condition. Outbreaks are caused by immune carriers passing the disease on to naïve birds. Virus is shed through faecal or oral discharge into food and water. Changing hours of daylight and the onset of the breeding season may stimulate the shedding of this herpes virus, which most commonly arrives in spring.

**Symptoms:** Signs of the disease are listless birds with drooping wings, diarrhoea causing a matting of the vent feathers, and a dirty appearance of the head from increased secretions from the eyes and nostrils. Depending on the stage of the disease, haemorrhages or necrotic tissue are found from the oesophagus right through to the cloaca.

**Treatment:** There is no treatment for affected birds. Rapid vaccination of the whole flock will save some, but a supply may have to be imported from Intervet, Holland. To avoid this disease, exclude wild mallards which are the main disease vector.

## Eye Problems

**Cause:** Insufficient water for washing; bacterial infection.

**Symptoms:** Geese occasionally get a bacterial eye infection. The eye looks sore and runny, and they rub it against their plumage.

## DROPPED TONGUE

*The bird appears to have another 'beak' below the lower mandible.*

*The floor of the mouth has been cleaned out, and the loose skin below the jaw stitched and drawn tighter under a local anaesthetic at the vet's.*

*The lack of blood supply causes the skin to become necrotic, and it falls off after two to three weeks.*

**Treatment:** Treat the eye with an antibiotic cream at night when there is no access to water. Allow the bird to bathe freely in the day time. Toulouse are rather prone to slight foaming of the eye, particularly in spring. This condition may not respond to antibiotic treatment; no pathogen was found in a sample sent to the pathology laboratory. The condition may be due to physical irritation, for example mud, mites or a poor diet. Ensure that the bird is not infested with northern mite. A full, deep bucket of water containing two to three drops of Milton can be also beneficial for washing the eyes; encourage this by dropping a few grains of wheat into the bucket.

## Goose Parvovirus (GPV): *see* Derzsy's Disease

## Lead Poisoning

Lead shot may be ingested: four to five pellets will kill smaller birds, which become emaciated. Paralysis of the muscles will cause functional failure of the muscles of the gizzard (*see* Owen, 1980).

## 'Limberneck': *see* Botulism

## Listeria: *see* Coordination Loss

## Maggots

During spells of hot, damp summer weather when flies proliferate, eggs may be laid around the vent of a goose. This is more likely to happen with Toulouse than other breeds, which stay cleaner. Birds with access to plenty of bathing water will not suffer. The flies are more likely to settle on a bird which is already ill, or has been injured. Pick the maggots off, and use antibiotic spray (POM) if the skin is raw. The spray forms a protective skin.

## Mineral Deficiencies

Deficiencies are unlikely to occur if birds get a mixed diet. Calcium and phosphorus are added to poultry pellets in the correct ratio for growers, maintenance and layers rations. Extra calcium should always be made available in the breeding season as mixed poultry grit so that the females can help themselves to the amount they want for egg formation. Excess calcium is not necessary for goslings: the calcium content of layers pellets is higher than they need and can be harmful as it will then increase their demand for phosphorus. Growers pellets contain the correct amounts of these minerals.

Other minerals such as magnesium, manganese, copper, iron, iodine, potassium sodium, selenium and zinc are needed in trace amounts, and the feed label on the bags of pellets will indicate if these have been added. If the goslings are suffering recurrent problems with wobbly legs and slipped tendons despite a good diet, and it happens with more than one pair of geese, it would be worthwhile checking if the soils in your area are deficient in particular minerals. Deficiency in certain minerals as well as vitamins can result in leg problems. Farmers are often aware of local deficiencies because of sheep and cattle nutrition, and agricultural services may be able to give advice. Note that inbreeding also results in problems with legs and tendons.

## Mites: Red Mite and Northern Mite

Red mite commonly infests chicken houses but can get in the goose house. These mites live in the cracks of the woodwork in the daytime, and suck the blood of the bird at night. They can live for months without food, when they are grey. They look virtually the same as northern mite in that they are both arachnids, and are red when they have eaten. If the birds are scratching for no apparent reason, examine the joints of the shed and under any roofing felt for signs of mite, identified as red clumps. Treat the shed with a permethrin insecticide, or a product such as Duramitex Plus, which attacks the mites and eggs. This infestation is rare with waterfowl.

Northern mite is a red, blood-sucking mite which conducts its entire life cycle on the bird in as little as seven days. These mites die without a warm host. In waterfowl these mites seem only to infest the head and neck (occasionally the vent), where they cannot be preened away. Mites cause the bird to scratch and may be a contributory factor in eye-foaming in the Toulouse. It is difficult to keep birds completely free of the parasite as it is transmitted quickly from bird to bird. It is probably acquired at bird shows, and from wild birds.

The parasite is difficult to see on coloured birds; it is more easily discovered on white birds if the feathers are parted on the crown of the head. If birds are scratching, they should be treated with a permethrin powder (synthetic pyrethrum) by ruffling up the feathers when the bird is dry, preferably when it is shut in for the night. The process should be repeated seven to ten days later in order to kill mites from any eggs which have hatched.

If a problem persists, PHARMAQ ivermectin, very concentrated at 1.0 per cent, may be prescribed by a vet. This is a systemic 'spot on' formula, applied to the skin. It is not licensed for food chain poultry, but is licensed for parrots and pigeons in accordance with the Small Animal Exemption Scheme.

## Newcastle Disease

A highly infectious viral disease causing respiratory problems, diarrhoea and high mortality. It occasionally visits us from the continent – for example in 1996–7, probably carried by migratory starlings. It seemed to affect large, commercial establishments rather than places with a few birds. When a case in reported, zones of exclusion are drawn, and poultry movement halted.

This is a notifiable disease in the UK, with compulsory slaughter for highly pathogenic strains. A vaccine is available.

## Pasteurellosis (Fowl Cholera)

**Cause:** The bacterium *Pasteurella multocida* causes respiratory infection in domestic and wild birds; enteritis is also associated with the condition. It can be spread by wild birds, rats and mice, by nasal exudate as well as in the droppings. It occurs most often in autumn and winter, though it may occur at any time of the year.

**Symptoms:** The bird may have watery green droppings, difficulty in breathing and discharge from the mouth and nostrils, and is unable to get up. There is a loss of appetite, increased thirst and a high body temperature. In the later stages it loses control of the neck muscles.

**Treatment:** Isolate sick birds. Treat with a soluble antibiotic prescribed by the vet; if caught very early the bird may be saved. If it will not drink, inject it instead with a suitable antibiotic; also use a syringe (no needle) to make sure that the bird gets liquid down the throat and does not get dehydrated. Take care with oral administration that liquid does not enter the trachea and lungs.

Laing considers that the environment remains infected and that healthy birds may remain as carriers, so improving hygiene after the occurrence of the disease is important. Send infected carcases for incineration. Lime the affected ground and rest it.

## Pseudotuberculosis (Yersinia)

In dirty conditions, birds can develop internal white growths in the liver which resemble TB. Infected birds are weak, thin, may have diarrhoea and are infectious. As with many diseases, this cannot be readily diagnosed except by post mortem. Antibiotics can help. It can affect humans, though this is uncommon.

## Pneumonia

Pneumonia is supposed to be common in goslings up to six months of age, whilst they have fluffy backs – that is, lack of protection by contour feathers. Avoid problems by keeping goslings warm and dry at night, and on clean bedding. We have only ever had one case in an adult bird. Symptoms are laboured breathing and lack of appetite. This is difficult to cure with antibiotics unless caught early. Make sure that sheds are well ventilated.

## Reproductive System Diseases

Oviduct problems are best avoided by keeping geese fit, not fat. They should be allowed to lay in their normal annual cycle (not forced by artificial lighting). It is also important to select birds of average weight for their breed, and which lay eggs of average size; forcing up size by selection tends to produce birds which lay double-yolk eggs, and such birds are more likely to have oviduct problems. Birds suffering from acute oviduct problems in spring can die quite quickly for no obvious external reason. In chronic cases death may be delayed for several days or even weeks. Birds obviously suffering from a chronic, untreatable problem are best put down.

### *Egg Peritonitis*

When an egg is released from the ovary, it can fail to enter the oviduct and enters the body cavity instead. The peritoneum – the membrane that lines the abdominal cavity – can become inflamed due to the presence of egg yolk. The abdomen feels hot and swollen. In small quantities, the yolk material can be resorbed by the peritoneum. Quite often, peritonitis results from bacterial infection. Birds in lay should always be treated quietly to avoid such problems.

### *Impaction of the Oviduct*

The tube carrying the eggs from the ovaries to the cloaca may become impacted with cheesy material, and even broken eggshell, and may become infected.

### *Egg Binding*

A bird may sometimes have extreme difficulty in passing an egg. If this is low down in the oviduct and the eggshell can even be seen as the bird strains to pass it, the bird can be helped. Keep it warm at 25°C and give liquid calcium as a drench, if available. Squirt warm olive oil into the vent with a syringe (no needle), and try to aid the passage of the egg by holding it from the outside of the body (beneath the abdomen) and gently easing it outwards. In a bad case the muscles are tight and will not allow the egg's passage. A calcium injection and oxytocin may help to relax the muscles: ask the vet; in a worst case scenario the egg may have to be broken and the shell extracted by forceps. This is best done at the vet's, as any tearing will cause infection. Always give the bird a course of antibiotic to prevent infection of the oviduct.

*Prolapse*

The lower part of the oviduct may protrude from the vent (prolapse) as a result of egg binding or straining. This can be gently cleaned and replaced in some cases, but is likely to become a persistent problem. It may be best to put the bird down if the prolapse is bad, or if the problem recurs. Encouraging a bird to go broody, and to stop laying, can allow healing.

## Subcutaneous Emphysema

Otherwise known as 'air intake beneath the skin'. There are eight major air sacs in the respiratory system of most species of bird. If an air sac ruptures, air escapes internally and is trapped under the skin. The condition caused two of our Toulouse geese literally to inflate. The vet had previously seen the condition in pigeons: in one bird it was confined to the head region and neck, in the other it affected the whole body. Opinion is divided on how best to treat this: to release the air by puncturing, or to allow the pressure to build up and the damage repair itself. Both of our birds had to be put down; the initial trauma was unknown in these two cases.

## Sinus Infection

This condition is much rarer in geese than ducks. If a swelling develops at the base of the bill so that the bird's 'cheek' swells out, do not assume it is a tumour, but first treat with an appropriate antibiotic from the vet. The rather hard swelling should soften as it responds to the antibiotic, and after about a week should have gone. Watch the bird to check that infection does not recur in susceptible individuals.

## Staphylococcus Infection

Staphylococcus infection occurs as a localized infection of the tendon sheaths and at joints in the legs, and in the feet. The affected area is hot to the touch and swells; the bird limps. Infection can follow as a result of sprain injury or skin damage. Treat the bird with antibiotic injections. Infection is less likely in adults than in young birds. Avoid problems by having clean grazing, and feeding the birds well with pelleted feed, with extra vitamins and minerals, rather than just wheat and grass.

## Tumours

Tumours can arise in quite young geese, even those of two or three years old. They can develop on the wings, head and neck, and may bleed, often so that the bird looks in a worse mess than it actually is. Nevertheless the bird should eventually be put down.

## Vitamin Deficiencies

Vitamin deficiencies are unlikely to develop where birds are given some pelleted foods and kept in natural conditions – that is, grazing on clean pasture. As with humans, a mixed diet including fresh food is best. Green foods supply vitamins A and K; sunlight allows the bird to utilize vitamin D; B vitamins and vitamin E are provided by cereal grains. Fish meal and oils, added to starter and grower pellets, are especially important sources of vitamins. Deficiencies are most likely to occur in intensively reared goslings kept indoors and reared on starter rations designed for poultry.

Leg weakness is the most likely manifestation of vitamin deficiency. It may show as weak hocks in one- to four-week-olds, or wobbly legs leading to difficulty in balance. If goslings, perhaps of a breed or strain with a particularly high vitamin demand, are not given greens in the first two weeks they can develop curved toes, bowed legs and slipped tendons. These conditions cannot be cured and so the problem is best avoided. Adding niacin or brewer's yeast to the diet prevents the condition, or simply getting the goslings out on to grass.

## Wing Problems

Problems include slipped, oar and angel wing (*see* Chapter 20), which occur when the gosling is between four and five weeks old, when the blood-filled quills are sometimes too heavy to be supported by the final wing joint. It may happen in some seasons but not in others, in goslings bred from the same breeding pair; its occurrence is therefore more often dictated by environmental factors – such as rapid growth from too much protein in the diet in domesticated birds – rather than genetic disposition. *See* the section in Chapter 20 on feeding goslings; also treatment must be immediate to stop permanent deformity.

## Worms

*Caecal Worms (*Hetarakis gallinarum*)*

Caecal worms are generally harmless unless they occur in large numbers.

*Acuaria*

Humphreys (1975) refers to this particular roundworm as being a problem in waterfowl. The worm damages the proventriculus so that the organ ceases to work, and the bird becomes thin and dies. This worm has a stage in its life cycle when it lives in the body of the water flea Daphnia, which is common in ponds. A good strong flow of water is recommended as it stops the Daphnia from multiplying.

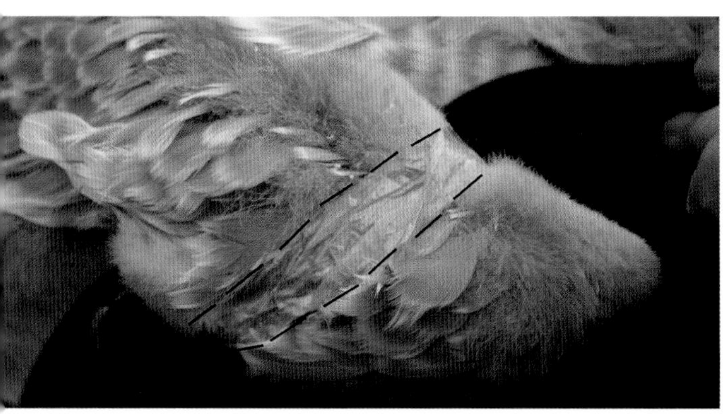

*Correcting angel wing: hold the wing in the correct position, and lightly tape it with Sellotape. The tape must not go into the crook of the joint because it will nip the flesh. Remove after two days maximum: growth is rapid, and the tape will tighten. Repeat if necessary.*

## Gape Worm

Worms in the trachea (*Cyathostoma bronchialis*) cause a bird to gasp for breath and 'sneeze'. The effect is similar to *Syngamus trachea* in poultry. Gape worms are less of a problem in geese than in chickens and ducks, but have been known to kill geese that were not treated. The infestation is detected by birds shaking their heads and rattling in their throat because they have difficulty in breathing; otherwise they behave normally and do not seem ill.

Infection is by the oral route in chickens, with earthworms, slugs and snails acting as transfer hosts. However, the life cycle may also be direct. The eggs of the gapeworm *Cyathostoma* (the gapeworm of geese and ducks) can be picked up directly from the grass. Some birds seem more prone to it than others, possibly because of eating soil to aid digestion. Ground that is 'poultry sick' is best rested and limed. Note that

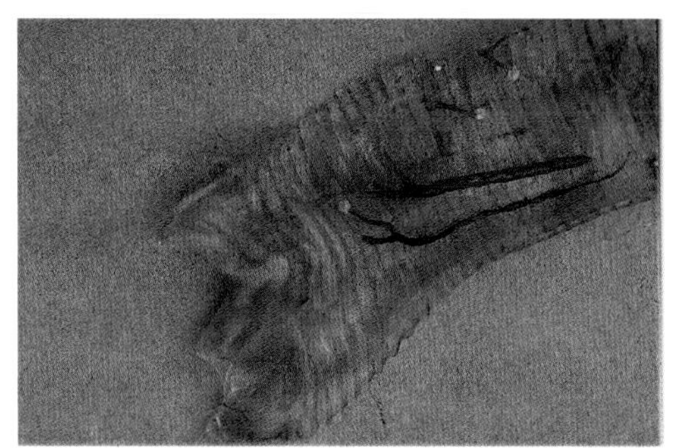

*Gapeworms: the trachea is opened out to show the characteristic Y-shape of both male and female. (Copyright Janssen Animal Health)*

infective larvae can become encysted within the bodies of invertebrate hosts, and can remain infective for more than a year.

## Gizzard Worm (Amidostomum anseris)

Gizzard worm is a very harmful worm for goslings. Like the gape worm, this nematode worm is spread directly on to the ground via the birds' droppings. The eggs hatch into larvae in one to three days, and are swallowed with the grass. The hair-like worms burrow under the horny lining of the gizzard and suck blood, causing ulceration and even rupture of the muscle so that the gosling cannot gain nutrition. The grinding pads of the gizzard are separated and become ineffective. Large numbers of worms are sometimes found in the gizzard of adults which have died from other causes. Adults may carry an infestation with no obvious effects; for goslings, however, a heavy infestation is fatal: they become emaciated, lethargic, fail to eat, and die.

Prevention is better than cure. The life cycle from egg to adult worm varies between three to five weeks, so worming and moving birds on to fresh pasture helps

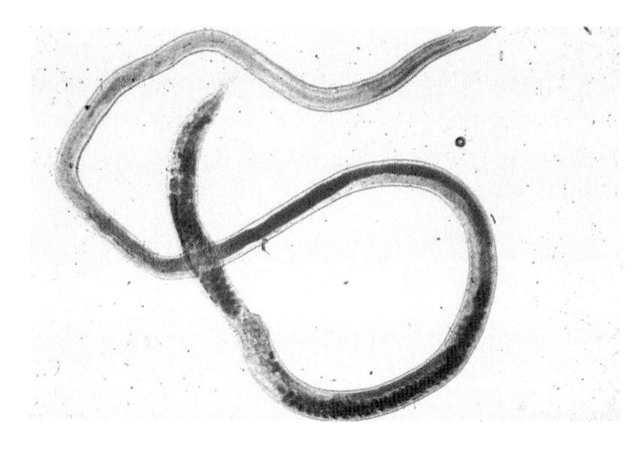

*Adult gizzard worms are ⅜–1.5in (10–35mm) long and usually bright red. (Copyright Janssen Animal Health)*

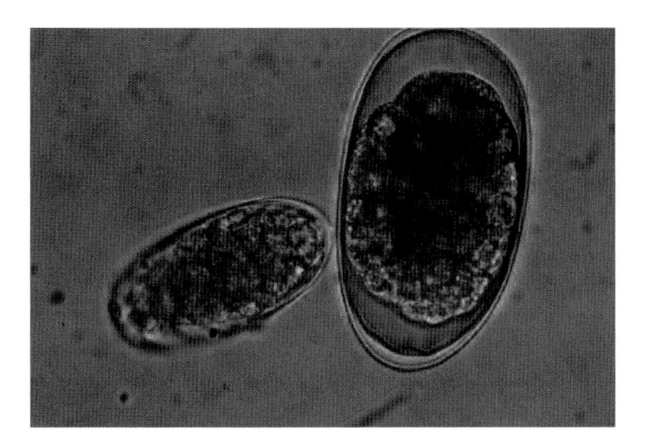

*Trichostrongylus and Amidostomum eggs. (Copyright Janssen Animal Health)*

break the cycle. Keeping grass short and dry also helps control the worm, which survives low temperatures at the infective larval stage, but not drying.

### Worming Geese

Worming should be done strategically: that is, the use of anthelmintics should be limited to key times and vulnerable individuals, so that resistance does not build up through overuse (*see* Chapter 20). Worms that are resistant survive and pass on their 'resistance' genes, and these worms accumulate until finally treatment failure occurs. It is therefore useful to change the wormer from time to time, and to use good practice to avoid the build-up of parasites.

**Flubendazole:** The approved wormer for geese, found in the product Flubenvet. Follow the dosage on the label. The Flubenvet powder is mixed with dry food, preferably pellets, and is fed to the birds for seven days. When a mixture of whole wheat and pellets is needed, the powder can be made to stick by first dressing the food mixture with 1 dessertspoonful of cooking oil per 7-13lb (3-6kg). Flubenvet powder can be prescribed by the vet or bought from retail outlets.

This is not a suitable wormer for geese when their main, and preferred, diet is grass, because insufficient worming powder will be consumed from the prepared food ration. Also, when a bird is really ill, it does not eat. It is possible to administer flubendazole as a drench (Solubenol); ask the vet for details.

**Panacur 10% solution:** This wormer is now licenced for commercial poultry and is used by vets. It will kill gizzard worm and gape worm but not their eggs. It can be used with adults and young goslings, but a second dose two weeks later is required to kill off any more parasitic worms which have developed from the egg stage in the meantime. The white liquid can be administered with care as a drench (see page 118). It can also be carefully dropped onto and mixed into pelleted food if this is evenly consumed.

**Prevention:** Problems with worms are best avoided by good management. Goslings should always be grazed on short, clean grass. If they are accompanied by adult geese, the adults should be wormed first. This is preferably done when the goose sits so that if she gets ill or too light, gizzard worm may be discounted as the problem.

The gizzard worm cycle is three to five weeks from egg to adult, so if the goslings have not been allowed on grass until two weeks old, there should be no initial problem.

Always keep a check on goslings by picking them up frequently to check their weight. Any bird which seems slow or sluggish should be examined and wormed.

Goslings rarely fall ill if they are fed crumbs, then pellets with wheat introduced ad lib. Those that are expected to survive largely on grass are more likely to suffer.

Note that **Piperazine**, which has been recommended as a wormer in the past, is not effective for goose parasites. Levamisole, found in sheep wormers, is no longer available because the parasites developed resistance. The same is true of goose parasites, and levamisole is now useless. **Ivermectin** has been used for waterfowl in recent years as a systemic treatment for all internal and external parasites. It is not licensed for poultry. It may be administered in tiny amounts on the skin, or injected, depending on the preparation. Ask your vet for details.

Apart from Flubenvet, these wormers are not approved for food chain birds: appropriate withdrawal times should be observed if meat and eggs are used. Note that routine worming is not recommended in the breeding season (only after egg production, when the goose sits) as these anthelmintics are toxic and can affect fertility.

***The effectiveness of various wormers on parasites affecting geese***

| *Wormer* | *Amidostomum* | *Syngamus* | *Heterakis* |
|---|---|---|---|
| Fenbendazole found in Panacur | No, eggs not killed | Yes | Yes |
| Flubendazole | Yes | Yes | Yes |
| Piperazine | No | No | Partly |

# Appendix
# Punnet Squares – American Buff × Embden

| **Parents** | **Spot** | **Dilution** | **Buff** | |
|---|---|---|---|---|
| American Buff ♂ | $Sp^+$ $Sp^+$ | $sd^+$ ($sd^+$) | g (g) | |
| Embden ♀ | sp sp | Sd – | $G^+$ – | |
| **$F_1$** | | | | |
| ♂ | $Sp^+$ sp | Sd $sd^+$ | $G^+$ g | Patchy light grey and white with no obvious buff |
| ♀ | $Sp^+$ sp | $sd^+$ – | g – | Buff with probable white feathers, e.g. primaries and chin |

**$F_2$**

| PARENTS ♂<br>♀ ( $F^1$) | **$Sp^+$ Sd $G^+$** | **$Sp^+$ $sd^+$ g** | **sp Sd $G^+$** | **sp $sd^+$ g** |
|---|---|---|---|---|
| **$Sp^+$ $sd^+$ g**<br>♂ | | $Sp^+$ $Sp^+$ $sd^+$ $sd^+$ g g<br>American Buff | | |
| **sp $sd^+$ g**<br>♂ | | | | sp sp $sd^+$ $sd^+$<br>g g<br>Buff Back |
| **$Sp^+$ – –**<br>♀ | $Sp^+$ $Sp^+$ Sd –<br>$G^+$ –<br>Pilgrim | $Sp^+$ $Sp^+$ $sd^+$ – g –<br>American Buff | | |
| **sp – –**<br>♀ | | | sp sp Sd –<br>$G^+$ –<br>Embden | sp sp $sd^+$ –<br>g –<br>Buff Back |

The empty squares are all heterozygotes
Compiled by Mike Ashton

# References


Acheson, H.V.C., *Modern Goose Keeping* (Crosby, Lockwood and Son, 1954).

Albin, E., *A Natural History of the Birds* (three volumes: 1731, 1734, 1738).

Afton, A.D. and Paulus, S.L., 'Incubation and Brood Care', *Ecology and Management of Breeding Waterfowl* Batt *et al* (University of Minnesota Press, 1992, pp. 62–108).

Ambrose, *The Aylesbury Duck* (Buckinghamshire County Museum, 1991).

American Poultry Association, *American Standard of Perfection* (1905).

Appleyard, R., *Geese, Breeding, Rearing and General Management* (Poultry World Ltd. , 1933)

Ashton, M., 'African Geese', *Fancy Fowl UK* (July, 2008)

Ashton, M., 'Explaining the Sebastopol' *Waterfowl* (Summer, 2009)

Belon, Pierre, *L'Histoire de la nature des oyseaux* (1555)

Bhatnagar, M. K., *Cytogenetic studies of some avian species* Ph.D. dissertation, University of Guelph, 1968.

Boswell, Peter, *The Poultry Yard* (First American edition, Wiley and Putnam, New York 1841).

Bowie, S.H.U., Shetland's native farm animals – Shetland Geese *The Ark* (1989).

British Poultry Standards, Ed. Victoria Roberts (Fifth Edition, 1997).

The Poultry Club Standards, Ed. W.W. Broomhead (1910, 1922, 1923, 1926, 1930).

The Poultry Club Standards, Ed. Threlford (1901, 1905).

British Poultry Standards, (Fourth Edition, 1954).

British Waterfowl Standards, reprinted from *British Poultry Standards* (4th edition, Butterworth, 1982).

British Waterfowl Association, British Waterfowl Standards (BWA, 2008).

Brooke, M. and Birkhead, T., *The Cambridge Encyclopedia of Ornithology* (CUP, 1991).

Brown, Dr A.F.A., *The Incubation book* (Wheaton, 1979).

Brown, E., *Races of Domestic Poultry (1906).*

Brown, E., 'The Rehabilitation of the Goose' *The Illustrated Poultry Record* (December 1913).

Brown, E., *Poultry Breeding and Production* (Caxton, 1929).

Brown, E., *British Poultry Husbandry – Its Evolution and History, 1830–1930* (Chapman Halls Centenary Year, 1930).

de Bruin 'The Steinbacher Fighting Goose' *Avicultura International (1995).*

Carefoot, Clive, 'Creative Poultry Breeding' (1983).

Cato, Marcus Porcius *Roman Farm Management* (Ed. F.H. Belvoir, 1918).

Charnock Bradley, O., *The Structure of the Fowl*, (Oliver and Boyd, 1950, revised by T. Grahame).

Clayton, G.A. 'Common Duck' *Evolution of Domesticated Animals* (Ed I.L. Mason, Longman, 1984).

Cowan, P.J., 'The goose: an efficient converter of grass? A review' *World Poultry Science Journal* (36, pp. 112–116).

Deeming, D.C., *Nests, Birds and Incubation* (Brinsea Products Ltd 2002).

Delacour, J., *The Waterfowl of the World* Vol. 4 (*County Life* Limited, London, 1964).

Dixon, the Rev. Edmund Saul, *Ornamental and Domestic Poultry* (London, 1848).

Dmitriev N.G. and Ernst L.K. (ed), *Animal genetic resources of the USSR* (FAO, Rome, 1989).

Dürigen, Bruno, *Geflügelzucht. Hand- und Lehrbuch der Rassenkunde, Zucht, Pflege und Haltung von Haus-, Hof- und Parkgeflügel. Band 1: Arten und Rassen* (Parey Hamburg 1921).

FAO, Report on Domestic Animal Genetic Resources in China, June 2003 ftp://ftp.fao.org/docrep/fao/010/a1250e/annexes/CountryReports/China_E.pdf

Beijing ftp://ftp.fao.org/docrep/fao/010/a1250e/annexes/CountryReports/China_E.pdf

Fabricius, E., 'What makes plumage waterproof?' *Tenth Annual Report of the Wildfowl Trust, 1957–1958* (Bailey and Sons, 1959).

Girling, S.J., 'Feather Growth, including Nutritional and Endocrine Factors' (The Braid Veterinary Hospital, Edinburgh EH9 3AZ).

Gordon, C.D., 'Sexual dimorphism in the down colour and adult plumage of geese' *Journal of Heredity* (29: 335–337, 1938).

Grow, Oscar, *Modern Waterfowl Management* (American Bantam Association, 1972).

Gupta, Lokesh *A comprehensive practical guide to improving egg quality* www.thepoultrysite.com

Hale, Thomas, *A Compleat Body of Husbandry* (Vol. II, 1756).

Hawes, R.H., 'The Origin of Pilgrim Geese', *Fancy Fowl* (1996).

Hawes, R.H., *'Mutations and Major Variants in Geese'* (Ch. 18, *Poultry Breeding and Genetics* ed. R.D. Crawford, Elsevier, 1990).

Hewetson, J., letter to *The Feathered World* (1 Feb. 1929).

Hoffmann, Edmund, 'The Chinese origin of the so-called African goose' *Fancy Fowl*, (1991).

Humphreys, P.N., 'Worms in Waterfowl', *Waterfowl* (1975).

Hurst, J.W., *Utility Ducks and Geese* (Constable and Co. 1919).

Irvine, Richard *et al*, 'Goose Parvovirus in Great Britain' *Veterinary Record,* (11 October 2008, Volume 163, Issue 15).

Ives, P., *Domestic Geese and Ducks* (Orange Judd Publishing Co, Inc.1947).

Jenyns, Leonard, *A Manual of British Vertebrate Animals: Or Descriptions of All the Animals Belonging to the Classes, Mammalia, Aves, Reptilia, Amphibia, and Pisces* (1835).

Jerome, F. N. ,'Colour inheritance in geese and its application to goose breeding', *Poultry Science* (Vol. XXII, No. 1, 1953).

Jerome, F.N., 'Inheritance of plumage colour in Domestic Geese' *Proc. 14th World Poultry Congress* (Madrid) 2: 73-76 (1970).

Johnson, Lt Col. A.A., *Chinese Geese* (1950).

Krapu, G.L., Reinecke, K.J. 'Foraging Ecology and Nutrition', *Ecology and Management of Breeding Waterfowl,* Batt *et al* (University of Minnesota Press, 1992, pp. 62-108).

Llewellyn, R., 'The Brecon Buff Goose', *The Feathered World* (26 Oct. 1934).

Lorenz, Konrad, *The Year of the Greylag Goose* (Eyre Methuen, 1979).

Lorenz, Konrad *Here I am–where are you? The behaviour of the greylag goose* 1991.

Markham, Gervase, *The English Husbandman* (1613).

Mattocks, J.G., 'Goose feeding and cellulose digestion' *Wildfowl* (The Wildfowl Trust, Slimbridge, Vol. 22, 1971).

Ministry of Agriculture, Fisheries and Food *Ducks* (MAFF Publications PB0079, 1990).

Moubray, Bonington, *A practical treatise on breeding, rearing and fattening all kinds of Domestic Poultry, Pheasants, Pigeons and Rabbits*, Second Edition (Sherwood, Neely and Jones, London, 1816).

Moubray, Bonington, *A practical treatise on breeding, rearing and fattening all kinds of Domestic Poultry, Pheasants, Pigeons and Rabbits* Eighth Edition (Sherwood, Gilbert & Piper, London, 1842).

Nolan. J. J., *Ornamental, Aquatic and Domestic Fowl and Game Birds* (1850).

Oswald, Paul-Erwin, *Unsere Gänse und Enten*, Oertel & Spörer (2006).

Owen, M., *Wildfowl of Europe* (Macmillan, 1997).

Parkinson, Richard, *Treatise on the Breeding and Management of Livestock* (Cadell and Davies, Strand, and Scholey, Paternoster-Row, 1810).

Parr, J., 'The Embden Geese in Germany', *Waterfowl* (1996).

Perrier, Edmond; Salmon, Julien, *La Vie des Animaux Illustrée - Les Oiseaux* (1905).

Proctor, Noble S. and Lynch, Patrick, J. *Manual of Ornithology* (Yale University Press, 1993).

Ray, John, *The Ornithology of Francis Willughby* (1678).

Roberts, M. and V., *Domestic Duck and Geese in Colour* (Domestic Fowl Trust, 1986).

Robinson, J.H., *Popular Breeds of Domestic Poultry* Reliable Poultry Journal Pub. Co. (324-5, 1924).

Robinson, J. H., *The Growing of Ducks and Geese for Profit and Pleasure* (1924).

Roullier-Arnoult, *Instructions Pratiques sur L'Incubation et L'Elevage Artificiels des Volailles* Fourth Edition, Paris, Librairie Agricole de la Maison Rustique (1880).

Schmidt, H., letter in *The Magazine of Ducks and Geese* Vol. X, No. 11959.

Schmidt, H., *Puten, Perlhuhner, Ganse, Enten* (Neumann-Neudamm,1989).

Soames, Barbara, *Keeping Domestic Geese* (Blandford, 1980).

Silversides F.G., Crawford R.D., and Wang H.C., 1988 'The Cytogenetics of Domestic Geese' *The Journal of Heredity* (1988:79(1):6-8).

Shi X-W., Wang J-W., Zeng F-T. and Qiu X-P, 'Mitochondrial DNA Cleavage Patterns Distinguish Independent Origin of Chinese Domestic Geese and Western Domestic Geese', *Biochemical Genetics,* Springer Netherlands (44, Numbers 5-6, June pp. 237-245, 2006).

Stevenson, Mary H., 'Nutrition of Domestic Geese' *Proceedings of the Nutrition Society* (48, 1989).

Stoll, Andreas, 'Auto-sexing Geese' *Waterfowl* (Autumn, 1984).

Tegetmeier, W.B., *The Poultry Book* (Routledge and Sons, London, 1867).

Todd, F., *Waterfowl: Ducks, Geese and Swans of the World* (Sea World Press, 1979).

van Gink C.S.Th., *The Feathered World* (1931).

van Gink, C.S.Th., *Siergevogelte* (de Haan, Utrecht , 1941).

Voitellier, Charles, *Avicultura, Encyclopédie Agricol* (1918).

Weir, Harrison, *Our Poultry* (Hutchinson and Co., London, 1902).

Wingfield, Rev. W. and Johnson, G.W., *The Poultry Book* (1853).

Wright, Lewis, *The Illustrated Book of Poultry* (1873).

Wright, Lewis, *The New Book of Poultry* (1902).

ZHU Wen-qi, LI Hui-fang, SONG Wei-tao, XU Wen-juan,CHEN Kuan-wei, 'Analysis on Systematic Evolution of Chinese Native Light-type Goose Breeds Based on mtDNA D-oop Sequences' (Poultry Institute, Chinese Academy of Agricultural Sciences, Jiangsu Provincial Key Lab of Poultry Genetics. Breeding, Yangzhou, Jiangsu 225009, China, 2010).

# Useful Contacts

**Allen & Page**
Tel: 01362 822900
www.smallholderfeed.co.uk

**Bio-Medicine**
www.bio-medicine.org/medicine-news/Scientists-Establish-Link-Between-Foie-Gras-and-Disease-22841-1

**British Goose Producers**
www.britishgoose.org.uk

**Brinsea Products Ltd**
Tel: 0345 226 0120 (local rate) or (+44) 1934 417523
www.brinsea.co.uk

**Defra**
www.gov.uk/government/organisations/department-for-environment-food-rural-affairs
www.gov.uk/government/publications/poultry-including-game-birds-registration-rules-and-forms

**Country Smallholding Magazine**
www.countrysmallholding.com/home

**Egg Crafters Guild**
Tel: 01271 341652
www.eggcraftersguild.webs.com

**Fancy Fowl Magazine**
Tel: 01728 622030
www.fancyfowl.com

**Gardencraft**
Top class poultry sheds
Tel: 01766 513036
Fax: 01766 514364
www.gcraft.co.uk/list/poultry_units/index.html

**The Norfolk Feather Company**
www.norfolk-feather.co.uk

**Humane Slaughter Association**
The Old School
Brewhouse Hill
Wheathampstead
Hertfordshire
AL4 8AN
United Kingdom
Tel: 01582 831919
Fax: 01582 831414
Email: info@hsa.org.uk
www.hsa.org.uk

**Interhatch**
Tel: 01246 264646
www.interhatch.com

**E F Shingfield and Sons**
Frost Row Farm House
Watton Road
Hingham
Norfolk
NR9 4NW
Tel: 01953 850259

***Smallholder Magazine*, Kelsey Publishing**
Editor Liz Wright
liz.wright@kelsey.co.uk
www.shop.kelsey.co.uk

**Solway**
Suppliers of products suitable for the small back garden hobbyist with only a handful of birds, right up to fully commercial enterprises
www.solwayfeeders.com

# Index